CliffsNotes®

Geometry

Practice Pack

By David Alan Herzog

Wiley Publishing, Inc.

Published by:
Wiley Publishing, Inc.
111 River Street
Hoboken, NJ 07030-5774
www.wiley.com

Published by Wiley Publishing, Inc., Hoboken, New Jersey
Published simultaneously in Canada

Library of Congress Control Number: 2009933752

ISBN: 978-0-470-48869-0

Printed in the United States of America

10 9 8 7 6 5 4

For general information on our other products and services or to obtain technical support, please contact our Customer Care Department within the U.S. at (877) 762-2974, outside the U.S. at (317) 572-3993, or fax (317) 572-4002.

Wiley also publishes its books in a variety of electronic formats. Some content that appears in print may not be available in electronic books. For more information about Wiley products, please visit our web site at www.wiley.com.

WILEY

This book is dedicated to Francesco, Sebastian, and Gino Nicholas Bubba, Jakob and Alex Cherry, Hailee Foster, Rocio, Myles, Kira, Reese, and Zoe Herzog, Marcelo, all their parents, and, of course, Uncle Ian.

It is also dedicated to the memories of Henry and Margaret Smolinski, George and Rose Herzog, and to the memory of my late sister, Lois Herzog, who died much too young, and who never met a stray dog or cat for which she didn't try to find a good home.

About the Author

David Alan Herzog is the author of numerous books, most of which are in mathematics, and more than 100 educational software programs in several disciplines. He taught math education at Fairleigh Dickinson University and William Paterson College, was mathematics coordinator for New Jersey's Rockaway Township Public Schools, and taught math in the New York City School System.

Publisher's Acknowledgments

Editorial

Project Editor: Donna Wright

Acquisitions Editor: Greg Tubach

Technical Editors: Tom Page and Mary Jane Sterling

Composition

Project Coordinator: Kristie Rees

Indexer: BIM Indexing & Proofreading Services

Proofreader: Toni Settle

Wiley Publishing, Inc. Composition Services

Table of Contents

Pretest

Directions: Questions 1 through 132.

Circle the letter of the appropriate answer.

1. A single letter, V, may be used to represent which of the following?

a. a ray
b. a point
c. a line segment
d. a line

2. Which of these extends infinitely far in two directions?

a. a plane
b. a ray
c. a line
d. a line segment

3. Which of these is defined by a minimum of three points, one of which is outside the line formed by the other two?

a. a ray
b. a line
c. a line segment
d. a plane

4. If a point lies outside a line then which of the following must be true?

a. The point and the line may or may not be in the same plane.
b. The line must be in one plane and the point in a second one.
c. Exactly one plane contains the line and the point.
d. There may be up to three separate planes involved.

5. If two lines intersect, at how many points may they intersect?

a. one and only one
b. one or two
c. no more than three
d. an infinite number

6. What name is given to a portion of a line bounded by two endpoints?
   a. a ray
   b. a segment
   c. a section
   d. a sector

7. Which is the correct name of

   B A

   a. segment *BA*
   b. segment *AB*
   c. ray *AB*
   d. ray *BA*

8. Points *A*, *B*, *C*, *D*, *E*, *F*, and *G* are evenly spaced on the following segment. Which is the midpoint?

   $\overline{AG}$

   a. halfway between *C* and *D*
   b. *D*
   c. halfway between *D* and *E*
   d. *C*

9. Angle *FDG* is formed by the intersection of the following rays. Where must the vertex of the angle be?

   $\overrightarrow{DF}$ $\overrightarrow{DG}$

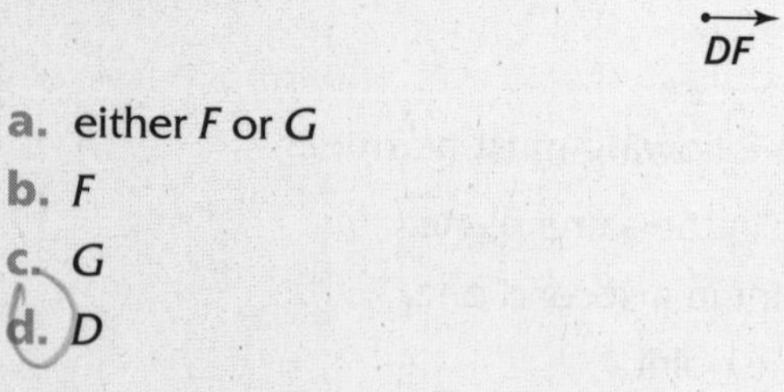

   a. either *F* or *G*
   b. *F*
   c. *G*
   d. *D*

10. Select the one correct letter-name for the angle formed by adding $\angle 3$ and $\angle 4$.

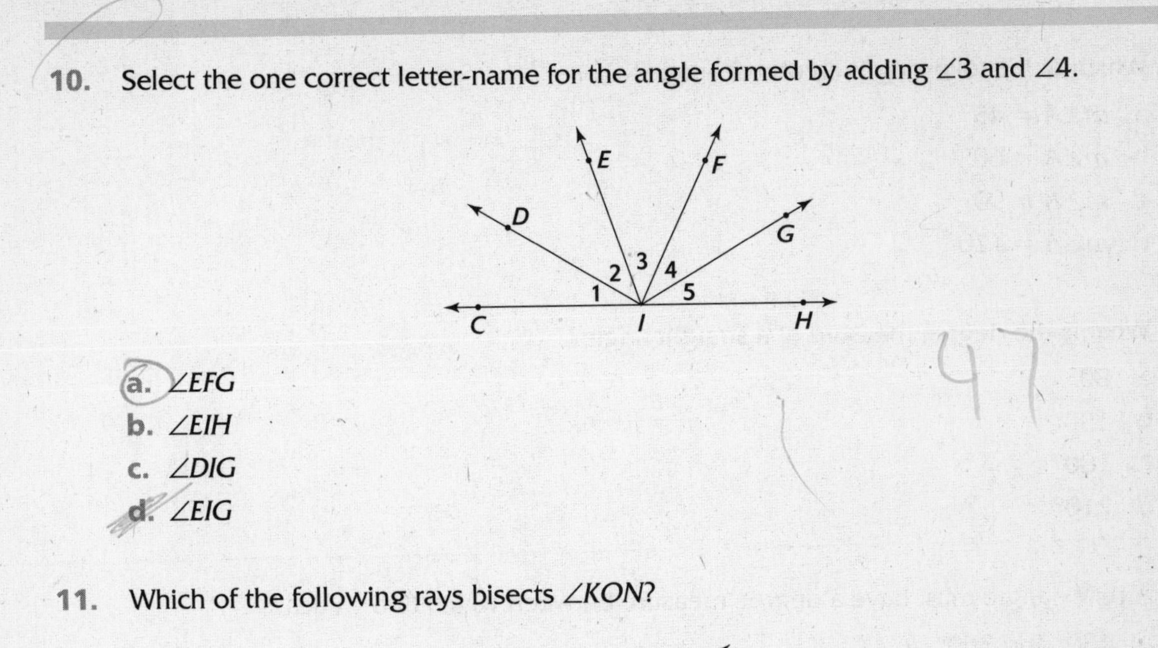

a. $\angle EFG$

b. $\angle EIH$

c. $\angle DIG$

d. $\angle EIG$

11. Which of the following rays bisects $\angle KON$?

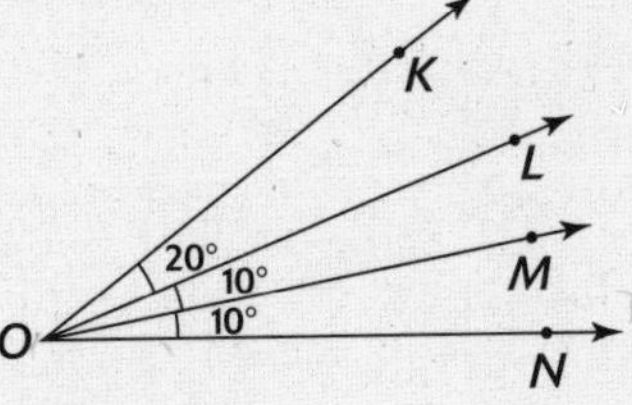

a. $\overrightarrow{OL}$

b. $\overrightarrow{OM}$

c. $\overrightarrow{LO}$

d. $\overrightarrow{KO}$

In the following three questions, "$m\angle A$" means the degree measure of angle A.

12. Which of the following describes a right angle?

a. $m\angle A = 45°$

b. $m\angle A = 60°$

c. $m\angle A = 90°$

d. $m\angle A = 120°$

13. Which of the following is not an acute angle?

a. $m\angle A = 45°$

b. $m\angle A = 60°$

c. $m\angle A = 75°$

d. $m\angle A = 100°$

14. Which of the following is an obtuse angle?

a. $m\angle A = 45°$

b. $m\angle A = 60°$

c. $m\angle A = 90°$

d. $m\angle A = 120°$

15. What is the degree measure of a straight angle?

a. 80°

b. 150°

c. 180°

d. 210°

16. A reflex angle must have a degree measure between which two values?

a. 180° and 360°

b. 170° and 270°

c. > 180° and < 270°

d. > 180° and < 360°

17. In the following figure, which answer-choice names a pair of adjacent angles?

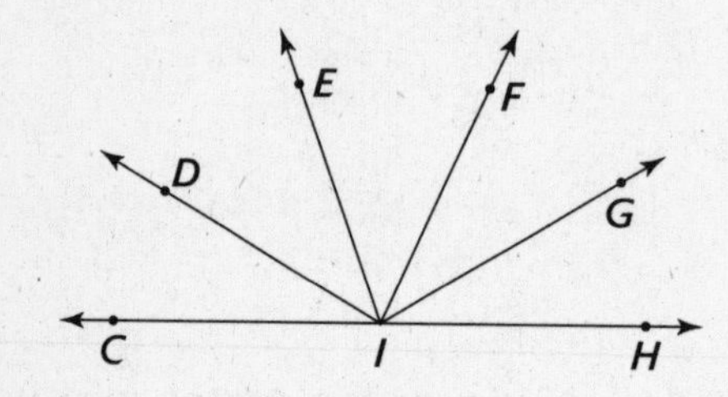

a. $\angle CID$ and $\angle DIC$

b. $\angle CID$ and $\angle DIE$

c. $\angle CID$ and $\angle EIF$

d. $\angle CID$ and $\angle GIH$

18. Where lines l and m cross in the following figure, which answer choice names a pair of vertical angles?

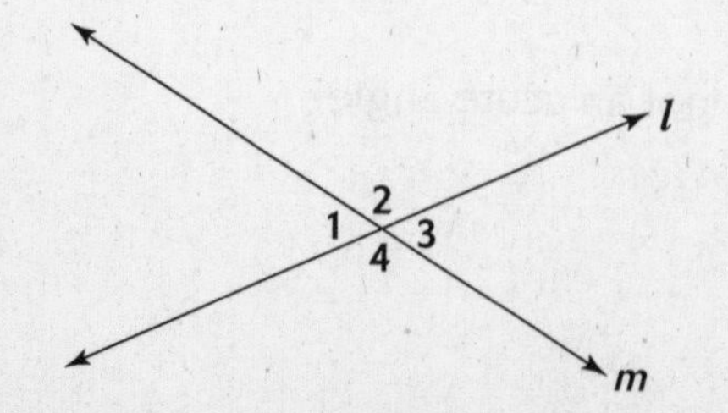

a. $\angle 1$ and $\angle 3$

b. $\angle 2$ and $\angle 3$

c. $\angle 4$ and $\angle 3$

d. $\angle 1$ and $\angle 4$

Use this diagram to answer questions 19 through 23.

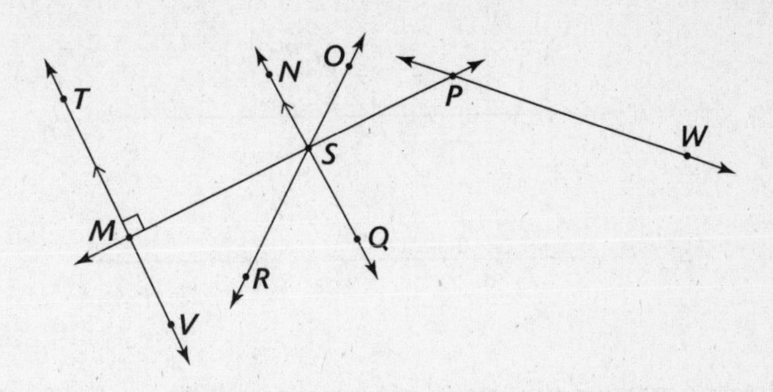

19. Which answer choice names a pair of complementary angles?

a. ∠*OSP* and ∠*OSQ*
b. ∠*OSP* and ∠*PSR*
c. ∠*MSN* and ∠*OSN*
d. ∠*MSR* and ∠*RSQ*

20. Which answer choice names a pair of supplementary angles?

a. ∠*OSP* and ∠*OSQ*
b. ∠*OSP* and ∠*PSR*
c. ∠*MSN* and ∠*OSN*
d. ∠*MSR* and ∠*RSQ*

21. Which answer choice does *not* name a pair of intersecting lines?

a. *PW* and *MS*
b. *NQ* and *MP*
c. *RO* and *MS*
d. *NQ* and *VT*

22. Which answer choice names a pair of perpendicular lines?

a. *PW* and *MS*
b. *NQ* and *MP*
c. *RO* and *MS*
d. *NQ* and *VT*

23. Which answer choice names a pair of parallel lines?

a. *PW* and *MS*
b. *NQ* and *MP*
c. *RO* and *MS*
d. *NQ* and *VT*

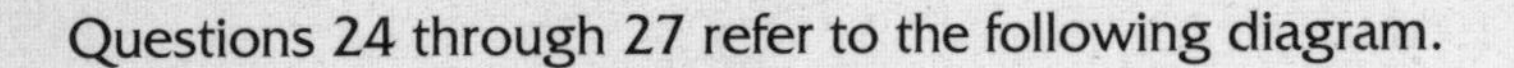

Questions 24 through 27 refer to the following diagram.

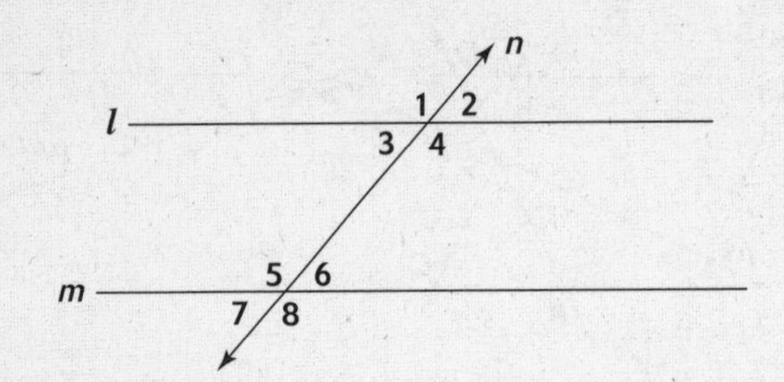

24. Which of the following names two pairs of corresponding angles for parallel segments *l* and *m*?
    - a. $\angle 1$, $\angle 5$ and $\angle 4$, $\angle 8$
    - b. $\angle 4$, $\angle 5$ and $\angle 3$, $\angle 6$
    - c. $\angle 7$, $\angle 2$ and $\angle 1$, $\angle 8$
    - d. $\angle 3$, $\angle 5$ and $\angle 4$, $\angle 6$

25. Which of the following names two pairs of alternate interior angles for parallel segments *l* and *m*?
    - a. $\angle 1$, $\angle 5$ and $\angle 4$, $\angle 8$
    - b. $\angle 4$, $\angle 5$ and $\angle 3$, $\angle 6$
    - c. $\angle 7$, $\angle 2$ and $\angle 1$, $\angle 8$
    - d. $\angle 3$, $\angle 5$ and $\angle 4$, $\angle 6$

26. Which of the following would serve to prove that segments *l* and *m* are parallel?
    - a. $\angle 1 = \angle 4$
    - b. $\angle 5 = \angle 8$
    - c. $\angle 5 = \angle 4$
    - d. $\angle 3 = \angle 2$

27. Which of the following would serve to prove that segments *l* and *m* are parallel?
    - a. $m\angle 1 + m\angle 2 = 180°$
    - b. $m\angle 4 + m\angle 6 = 180°$
    - c. $m\angle 7 + m\angle 8 = 180°$
    - d. $m\angle 5 + m\angle 7 = 180°$

28. In triangle *ABC*, $m\angle A = 45°$ and $m\angle B = 60°$. What must be the $m\angle C$?
    - a. 45°
    - b. 60°
    - c. 75°
    - d. 90°

29. What is the $m\angle G$ in $\triangle EFG$ (as follows)?

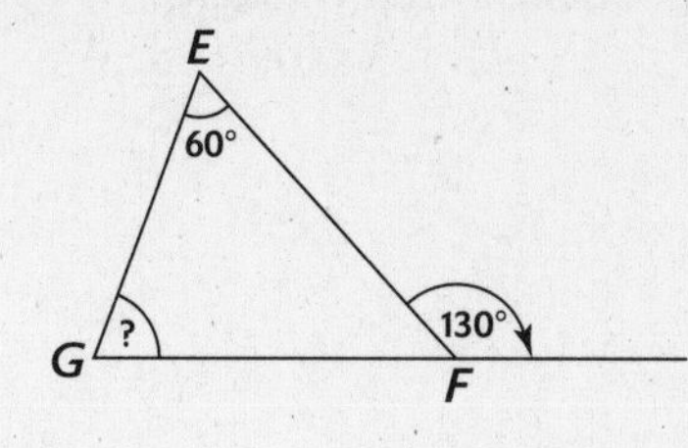

a. 50°
b. 60°
c. 70°
d. 80°

30. A triangle containing an angle larger than 90° is

a. an obtuse triangle.
b. a scalene triangle.
c. an acute triangle.
d. an impossibility.

31. What do we call a triangle in which two angles are of equal measure?

a. scalene
b. isosceles
c. equiangular
d. acute

32. What is the name given to a triangle with all three sides of different length?

a. acute
b. isosceles
c. scalene
d. There is no name for it.

33. In $\triangle KLM$, $m\angle M = 45°$, $m\angle L = 55°$, and $m\angle K = 80°$. Which is the longest side?

a. KL
b. LM
c. KM
d. can't tell

34. What name describes the two same-length sides of an isosceles triangle?

a. bases
b. altitudes
c. medians
d. legs

35. Which of the following sometimes is inside a triangle, sometimes coincides with a side of a triangle, and sometimes falls outside of a triangle?

a. the base
b. the altitude
c. the median
d. the angle bisector

36. Which three always meet at a single point inside the triangle?

a. the bases
b. the altitudes
c. the medians
d. the angle bisectors

37. What must be true of two triangles for them to be congruent?

I. They must be the same shape.
II. Their corresponding sides must be congruent.
III. They must be the same size.

a. I and II only
b. I and III only
c. II and III only
d. I, II, and III

38. Which of the following is a legitimate reason for declaring two triangles to be congruent?

a. SSA $\cong$ SSA
b. SAS $\cong$ SAS
c. AAA $\cong$ AAA
d. all of the above

39. Which of the following is a legitimate reason for declaring two triangles to be congruent?

a. ASA $\cong$ ASA
b. SAA $\cong$ SAA
c. HL $\cong$ HL
d. all of the above

40. The two triangles pictured here are congruent by SAA. Which of the following is true?

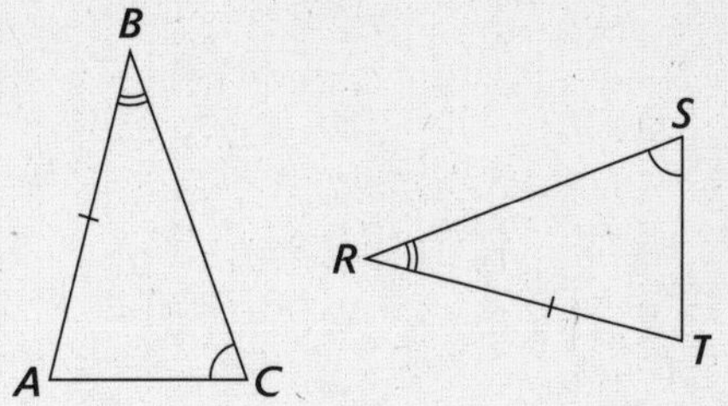

a. $RT \cong BC$
b. $\angle A \cong \angle R$
c. $BC \cong RS$
d. all of the above

41. Isosceles triangle *FGH* has congruent sides *FG* and *GH*. Which is its vertex angle?

a. $\angle F$
b. $\angle G$
c. $\angle H$
d. all of the above

42. Isosceles triangle *MNO* has base angles *M* and *O*. Which are its legs?

a. *MN* and *MO*
b. *MN* and *NO*
c. *MO* and *NO*
d. all of the above

43. In $\triangle ABC$, $\angle A = 30°$, $\angle B = 100°$, and $\angle C = 50°$. Which of the following lists the sides of the triangle in order of increasing size?

a. BC, AB, AC
b. AC, BC, AB
c. BC, AC, AB
d. AB, AC, BC

44. What is the minimum number of sides a figure can have in order to be classified as a polygon?

a. two
b. three
c. four
d. five

45. A concave polygon must have an interior angle with a degree measure larger than which amount?

a. 90°

b. 120°

c. 150°

d. 180°

46. A diagonal of a polygon is a line segment that connects which of these?

a. any two vertices

b. any opposite vertices

c. any two nonconsecutive sides

d. any two nonconsecutive vertices

47. An equilateral triangle is always equiangular, and an equiangular triangle is always equilateral. Is this true of other polygons?

a. sometimes

b. always

c. never

d. can't tell

48. What is the sum of the interior angles in a quadrilateral?

a. 180°

b. 360°

c. 540°

d. 720°

49. What is the sum of the interior angles in a regular hexagon?

a. 180°

b. 360°

c. 540°

d. 720°

50. What is the most appropriate name for a four-sided figure with one pair of sides parallel?

a. quadrilateral

b. trapezoid

c. rhombus

d. parallelogram

51. If one diagonal is drawn in a figure and results in two congruent triangles being formed, then the figure *must* be a _________.

a. rectangle
b. trapezoid
c. rhombus
d. parallelogram

52. A trapezoid's bases are 14 cm and 28 cm. What is the length of its median?

a. 18 cm
b. 21 cm
c. 24 cm
d. not enough information

53. Which of the following best describes a square?

I. a rhombus
II. a rectangle
III. a parallelogram

a. I and II together
b. I and III together
c. I, II, and III
d. II and III together

54. Which of the following conditions must be met to prove that a quadrilateral is a parallelogram?

a. Both pairs of opposite sides are equal.
b. Both pairs of opposite angles are equal.
c. One pair of opposite sides are equal and parallel.
d. all of the above

55. Which of the following conditions must be met to prove that a quadrilateral is a parallelogram?

a. The diagonals bisect each other.
b. The diagonals are equal.
c. It contains one right angle.
d. all of the above

56. Which quadrilateral's diagonals are perpendicular to each other and bisect the figure's opposite angles?

a. trapezoid
b. rectangle
c. rhombus
d. parallelogram

57. What is the minimum requirement for a parallelogram to be proved a rectangle?

a. one right angle
b. two right angles
c. three right angles
d. four right angles

58. Which would be a good description of a square?

I. an equilateral rhombus
II. an equiangular rhombus
III. an equilateral rectangle

a. I and II only
b. I and III only
c. II and III only
d. I, II, and III

59. Which properties are true in an isosceles trapezoid?

a. The base angles' degree measures are equal.
b. The diagonals are equal in length.
c. The nonparallel sides are equal in length.
d. all of the above

60. $\overline{FG}$ joins the midpoints of sides AB and AC of $\triangle ABC$. Side BC has a length of 12 units. What is true of $\overline{FG}$?

a. It has a length of 6 units and is parallel to AB.
b. It is parallel to AB, and its length cannot be determined from the given information.
c. It may or may not be parallel to AB, and its length is 7 units.
d. It may or may not be parallel to AB, and its length is 5 units.

61. A rectangle is 6 inches long and 4 inches wide. What is its perimeter?

a. 10 inches
b. 16 inches
c. 20 inches
d. 24 inches

62. A square is 8 cm wide. What is its area?

a. 32 cm^2
b. 64 cm^2
c. 128 cm^2
d. not enough information

Questions 63 and 64 refer to this figure:

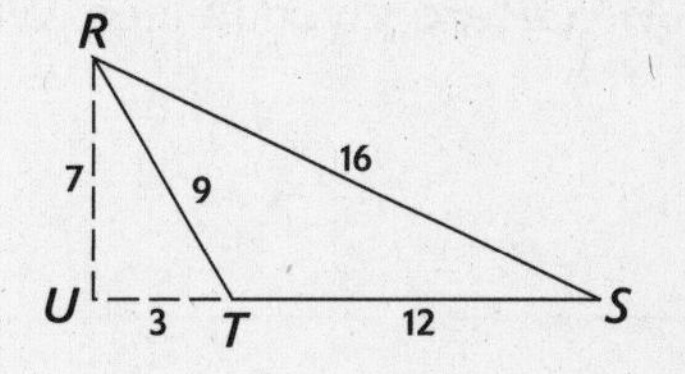

63. The numbers on $\triangle RST$ are in units of linear measure. Find the triangle's perimeter.

a. 19 units
b. 37 units
c. 38 units
d. 56 units

64. What is the area enclosed by $\triangle RST$?

a. 42 $units^2$
b. 73 $units^2$
c. 84 $units^2$
d. 112 $units^2$

65. What is the area of this parallelogram?

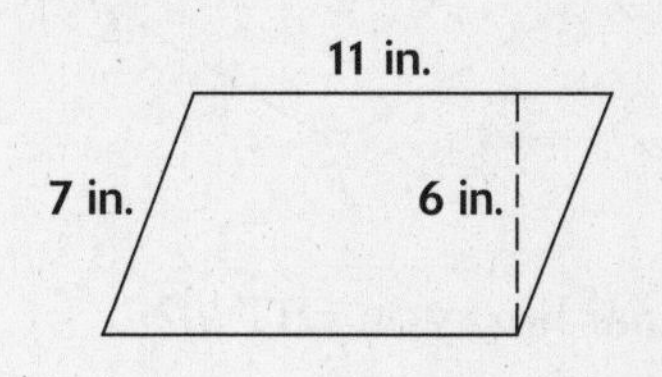

a. 36 in^2
b. 42 in^2
c. 66 in^2
d. 77 in^2

Questions 66 and 67 refer to this figure:

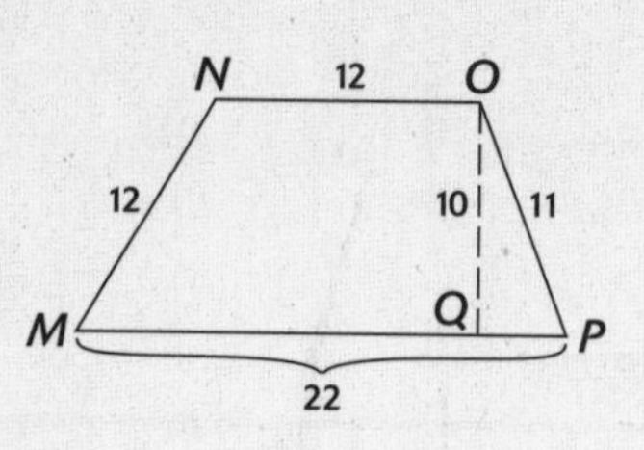

66. The dimensions of trapezoid *MNOP* are given in feet. What is the perimeter of the figure?

a. 56 ft.
b. 57 ft.
c. 120 ft.
d. 170 ft.

67. What is the area of trapezoid *MNOP*?

a. 120 ft^2
b. 144 ft^2
c. 170 ft^2
d. 340 ft^2

Questions 68 and 69 refer to this figure:

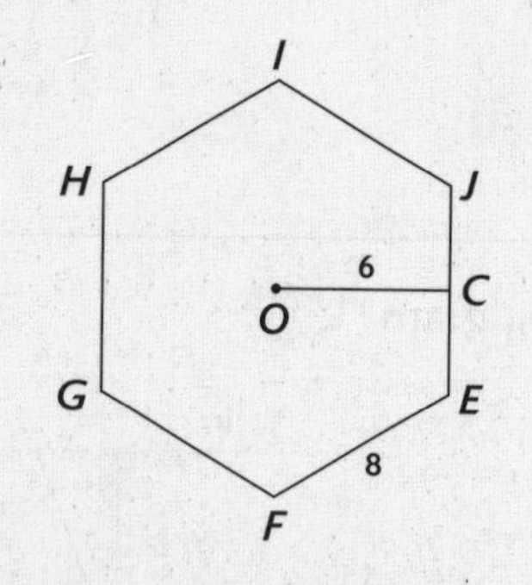

68. What is the perimeter of regular hexagon *EFGHIJ*?

a. 36 cm
b. 40 cm
c. 48 cm
d. 54 cm

69. What is the area of regular hexagon *EFGHIJ*?

a. 72 cm^2
b. 144 cm^2
c. 216 cm^2
d. 288 cm^2

Questions 70, 71, and 72 refer to this figure:

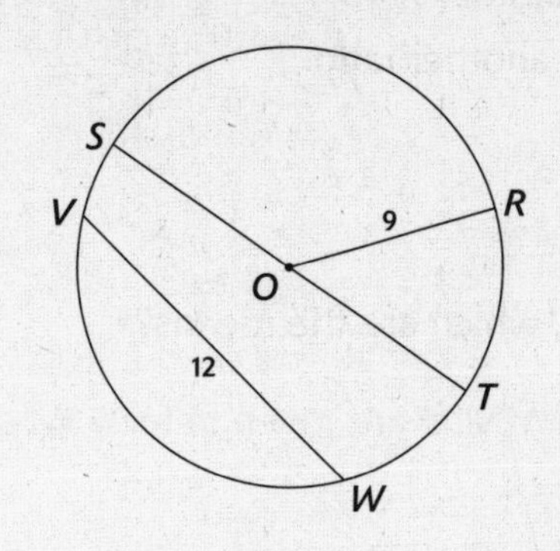

70. Which line segment in Circle O is its diameter?

a. $\overline{RO}$
b. $\overline{TO}$
c. $\overline{VW}$
d. $\overline{ST}$

71. What is the circumference of circle O?

a. 9π
b. 12π
c. 18π
d. 27π

72. Find the area of circle O?

a. 18π
b. 81π
c. 144π
d. 324π

73. A class contains 18 boys and 7 girls. What is the ratio of boys to children in the class?

a. 18:7
b. 7:18
c. 18:25
d. 25:18

74. The ratio of two supplementary angles is 2:3. What is the degree measure of the larger angle?

a. 36°
b. 54°
c. 72°
d. 108°

75. What does a proportion state?

a. One ratio is larger than another ratio.

b. One ratio is smaller than another ratio.

c. Two ratios are equal.

d. any of the above

76. In the proportion, $a:b = c:d$, which are the means?

a. b and c

b. a and d

c. b and d

d. a and c

Questions 77 and 78 refer to the following figure.

$$\frac{a}{b} = \frac{c}{d}$$

77. In the proportion, which are the extremes?

a. b and c

b. a and d

c. b and d

d. a and c

78. Which of the following is true of the proportion?

a. $ab = cd$

b. $ad = bc$

c. $ac = bd$

d. $ab = bc$

79. The scale on a map shows 5 cm is equal to 4 actual miles. If two towns on the map are 20 cm apart, how far apart are the actual towns?

a. 12 miles

b. 14 miles

c. 16 miles

d. 20 miles

80. What must be true of two similar polygons?

a. All the sides must be congruent.

b. All angles must be similar, and all sides must be in proportion.

c. All sides and angles must be similar.

d. All corresponding angles must be congruent, and all the ratios of corresponding sides must be equal.

81. How many corresponding angles must be proven to be equal in order to prove two triangles similar?

a. one

b. two

c. three

d. four

82. Two isosceles triangles have the same vertex angle. When would they be similar?

a. all of the time

b. some of the time

c. never

d. not enough information

Questions 83 and 84 refer to the following figure.

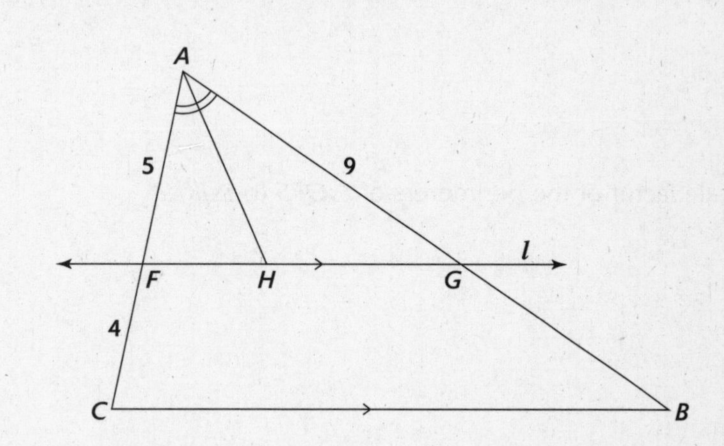

83. Line l intersects $\triangle ABC$ at F and G and is parallel to BC. The dimensions are as marked. What is the length of side AB?

a. 7.2

b. 9

c. 16.2

d. 18

84. AH bisects $\angle A$. Which of these is the ratio of FH to GH?

a. $\frac{5}{9}$

b. $\frac{5}{16}$

c. $\frac{7}{9}$

d. $\frac{9}{16}$

Questions 85, 86, and 87 refer to the following diagram.

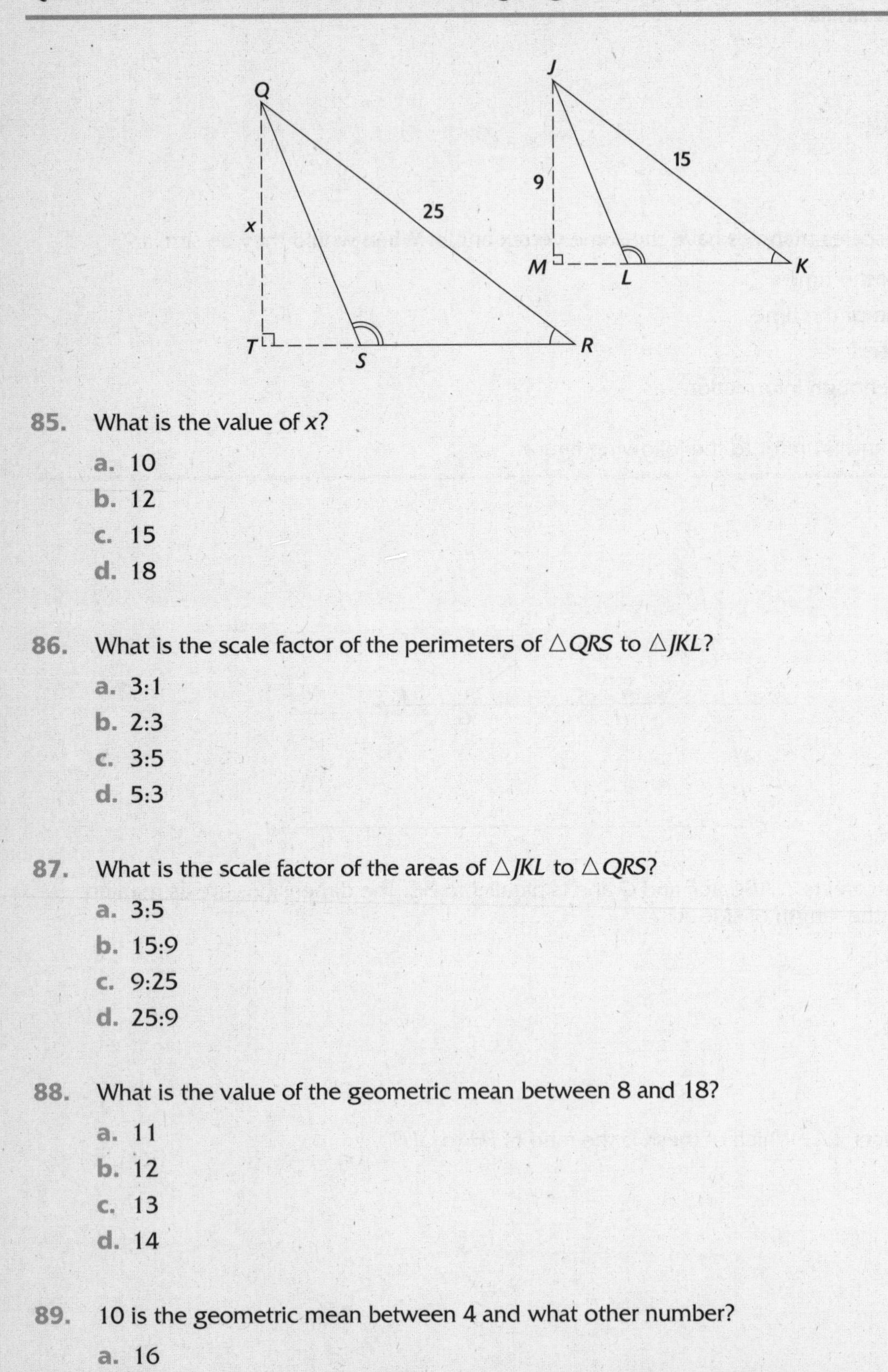

85. What is the value of x?

a. 10
b. 12
c. 15
d. 18

86. What is the scale factor of the perimeters of $\triangle QRS$ to $\triangle JKL$?

a. 3:1
b. 2:3
c. 3:5
d. 5:3

87. What is the scale factor of the areas of $\triangle JKL$ to $\triangle QRS$?

a. 3:5
b. 15:9
c. 9:25
d. 25:9

88. What is the value of the geometric mean between 8 and 18?

a. 11
b. 12
c. 13
d. 14

89. 10 is the geometric mean between 4 and what other number?

a. 16
b. 19
c. 22
d. 25

90. An altitude is drawn to the hypotenuse of a right triangle, dividing it into segments of 4 cm and 16 cm. What is the length of the altitude?

a. 6 cm
b. 8 cm
c. 10 cm
d. 12 cm

91. The legs of a right triangle are 6 and 8 inches long. How long is the hypotenuse?

a. 9 inches
b. 10 inches
c. 11 inches
d. 12 inches

92. The hypotenuse of a right triangle is 13 cm long. One of the legs is 5 inches long. How long is the other leg?

a. 9 cm
b. 10 cm
c. 11 cm
d. 12 cm

93. A triangle's sides are 5, 7, and 9 inches long. Therefore, the triangle is

a. acute
b. right
c. obtuse
d. can't tell

94. A triangle's sides are 6, 9, and 10 inches long. Therefore, the triangle is

a. acute
b. right
c. obtuse
d. can't tell

95. In a 30°-60°-90° right triangle, what's the length of the side opposite the 30° angle?

a. half the hypotenuse
b. half the hypotenuse $\sqrt{2}$
c. half the hypotenuse $\sqrt{3}$
d. not enough information

96. What is the name of a line segment whose endpoints are on the circumference of a circle?

a. radius
b. secant
c. chord
d. tangent

97. What line touches a circle's circumference at one point only?

a. radius
b. secant
c. chord
d. tangent

98. What segment touches a circle's circumference at exactly two points?

a. radius
b. secant
c. chord
d. tangent

99. How do we find the degree measure of an arc of a circle?

a. It equals the degree measure of its central angle.
b. It equals half the degree measure of its central angle.
c. It equals the degree measure of its inscribed angle.
d. none of these

100. What is the degree measure of a major arc of a circle?

a. $> 90^\circ$
b. $> 90^\circ$ and $< 180^\circ$
c. 180°
d. $> 180^\circ$

101. What is the degree measure of $\angle EFG$ in the diagram?

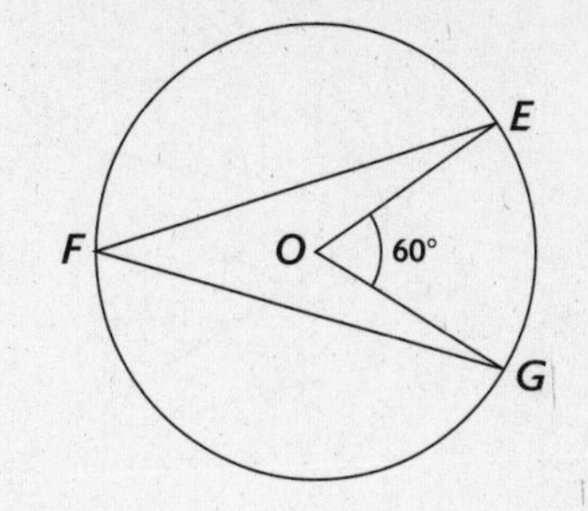

a. 20°
b. 30°
c. 40°
d. 50°

102. Which choice names the measure of the angle formed by radius MS and tangent QR at S?

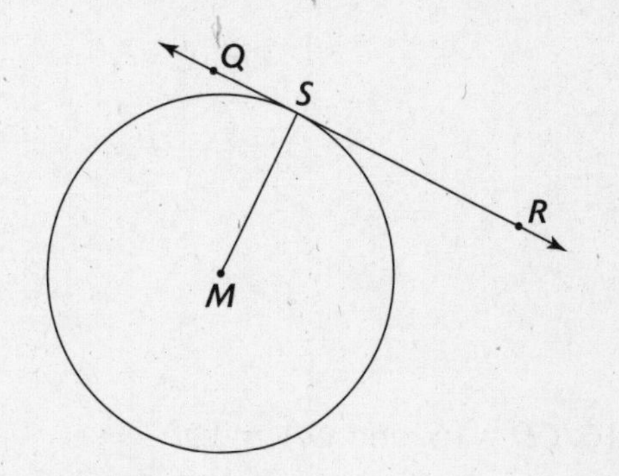

a. 80°
b. 90°
c. 100°
d. can't be sure

Refer to the following diagram to answer questions 103 and 104.

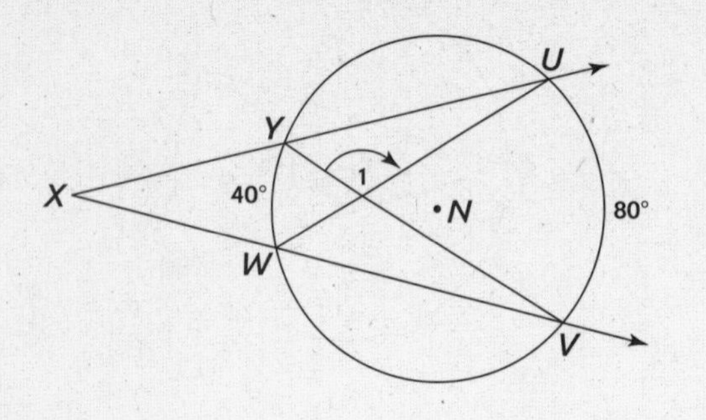

103. Find the measure of $\angle X$.

a. 20°
b. 30°
c. 40°
d. 60°

104. Find the measure of $\angle 1$.

a. 60°
b. 80°
c. 100°
d. 120°

In the following figure, $AB = 16$, $CD = 16$, and $OA = 10$.

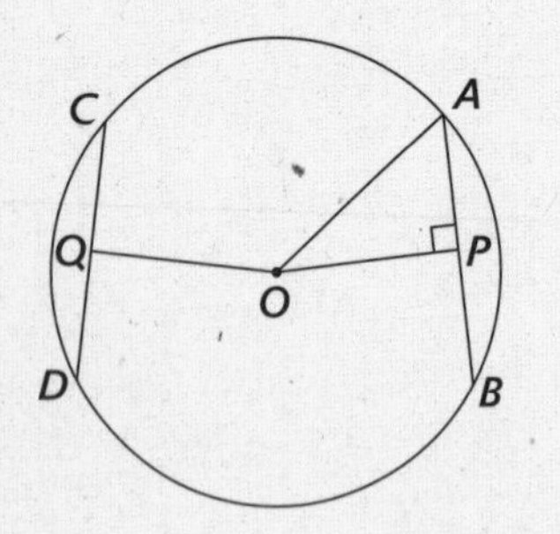

105. What is the length of OQ?

a. 6
b. 8
c. 10
d. 12

106. Find the length of x.

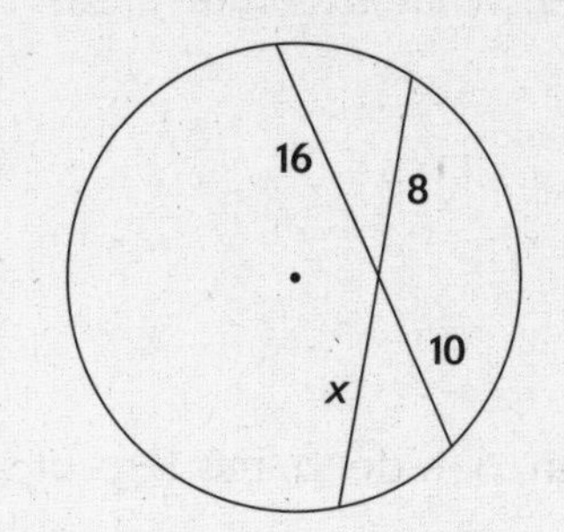

a. 12
b. 16
c. 20
d. 24

107. A circle has a radius of 10 cm. What is the length of an arc formed by a 60° central angle?

a. $\frac{10\pi}{6}$
b. $\frac{10\pi}{3}$
c. $\frac{10\pi}{2}$
d. 10π cm

108. A circle has a diameter of 12 cm. What is the area of a sector formed by a 120° central angle?

a. 4π cm^2
b. 8π cm^2
c. 12π cm^2
d. 24π cm^2

109. A right pentagonal prism has a height of 6 inches and sides of length 4, 5, 6, 5, and 7 inches, respectively. What is the prism's lateral area?

a. 120 in^2
b. 162 in^2
c. 284 in^2
d. 4200 in^2

110. A right triangular prism has a height of two cm and legs of 3 and 4 cm, respectively, that are perpendicular to each other. What is its total area?

a. 18 cm^2
b. 24 cm^2
c. 36 cm^2
d. 48 cm^2

111. A trapezoidal prism is 6 ft. high, 6 ft. deep, has legs of 10 ft. each and bases of 10 and 22 ft. What is its volume?

a. 96 ft^3
b. 192 ft^3
c. 252 ft^3
d. 576 ft^3

112. A cylindrical paint can is 10 inches high and has a radius of 4 inches. What is its lateral area?

a. 16π in^2
b. 32π in^2
c. 80π in^2
d. 112π in^2

113. Find the volume of that same paint can (from question 112).

a. 40π in^3
b. 80π in^3
c. 120π in^3
d. 160π in^3

114. What shape forms the base of a regular pyramid?

a. a triangle
b. a square
c. a rectangle
d. any polygon

115. A square pyramid's base is 16 meters long. It has a height of 9 meters and a slant height of 10 meters. What is its total area?

a. 320 m^2
b. 472 m^2
c. 576 m^2
d. 620 m^2

116. Find the volume of the pyramid in question 115.

a. 576 m^3
b. 768 m^3
c. 1440 m^3
d. 1728 m^3

A right circular cone has a height of 12 inches and a slant height of 15 inches. The diameter of the base is 16 inches. Questions 117, 118, and 119 refer to this hypothetical figure as described.

117. What is the cone's lateral area?

a. 104π in^2
b. 112π in^2
c. 120π in^2
d. 128π in^2

118. What is the cone's volume?

a. 64π in^3
b. 128π in^3
c. 256π in^3
d. 512π in^3

119. Find the cone's total area?

a. 120π in^2
b. 184π in^2
c. 196π in^2
d. 256π in^2

Questions 120 and 121 refer to a sphere with a 3-meter radius.

120. Find the volume of the sphere.

a. 12π m^3
b. 36π m^3
c. 72π m^3
d. 108π m^3

121. What is the sphere's surface area?

a. 18π m^2
b. 27π m^2
c. 36π m^2
d. 45π m^2

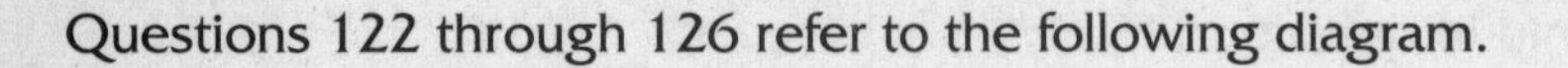

Questions 122 through 126 refer to the following diagram.

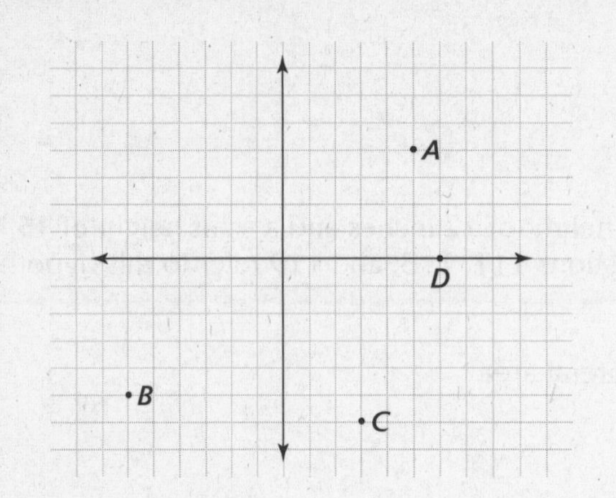

122. Which of the following names the coordinates of Point *A*?

a. (4, 5)
b. (5, 4)
c. (5, −4)
d. (4, −5)

123. Which of the following names the coordinates of Point *B*?

a. (6, −5)
b. (−5, −6)
c. (−6, −5)
d. (−5, −6)

124. Which of the following names the coordinates of Point *C*?

a. (6, −3)
b. (−6, −3)
c. (−3, 6)
d. (3, −6)

125. What is the distance between points *B* and *D*?

a. 5
b. 11
c. 12
d. 13

126. Find the coordinates of the midpoint of the line segment that if drawn would connec
points *A* and *C*.

a. (4, −1)
b. (−1, 4)
c. (5, −2)
d. (−2, 5)

Questions 127 through 130 refer to the following diagram.

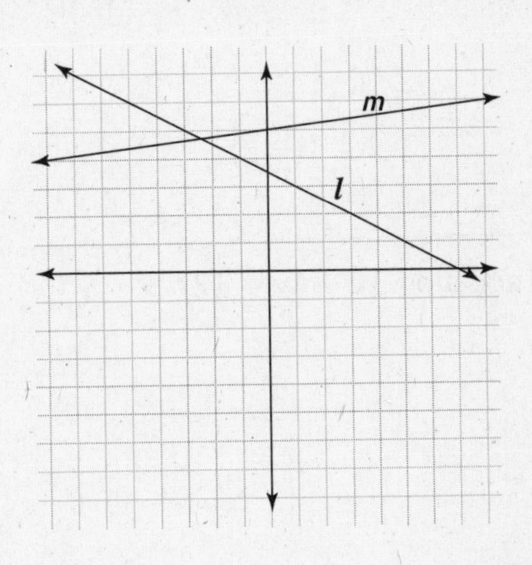

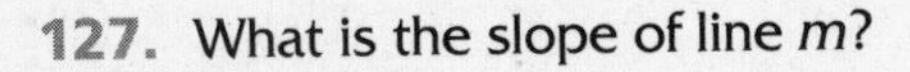

127. What is the slope of line *m*?

a. $-\frac{1}{8}$
b. $\frac{1}{8}$
c. −6
d. −8

128. What is the slope of line *l*?

a. $-\frac{1}{2}$
b. $\frac{1}{2}$
c. −2
d. −3

129. Line *n* (not shown) is parallel to line *m*. What can you conclude about its slope?

a. It is +8.
b. It is −8.
c. It is $+\frac{1}{8}$.
d. It is $-\frac{1}{8}$.

130. Line p (not shown) is perpendicular to line l. What can you conclude about its slope?

a. It is +2.

b. It is −2.

c. It is $+\frac{1}{2}$.

d. It is $-\frac{1}{2}$.

131. What is the y-intercept of the equation $3x - 4y = 12$?

a. −3

b. $\frac{3}{4}$

c. 3

d. $-\frac{3}{4}$

132. What is the slope of the equation $3x - 4y = 12$?

a. −3

b. $\frac{3}{4}$

c. 3

d. $-\frac{3}{4}$

Answers

1. b

2. c

3. d

If you missed 1, 2, or 3, go to "Naming Basic Forms," page 37.

4. c

5. a

If you missed 4 or 5, go to "Postulates and Theorems," page 38.

6. b

7. c

8. b

If you missed 6, 7, or 8, go to "Finding Segments, Midpoints, and Rays," page 41.

9. d

10. d

If you missed 9 or 10, go to "Forming and Naming Angles," page 45.

11. a

If you missed 11, go to "Angle Bisector," page 47.

12. c

If you missed 12, go to "Right Angles," page 47.

13. d

If you missed 13, go to "Acute Angles," page 48.

14. d

If you missed 14, go to "Obtuse Angles," page 48.

15. c

If you missed 15, go to "Straight Angles," page 48.

16. d

If you missed 16, go to "Reflex Angles," page 48.

17. b

If you missed 17, go to "Adjacent Angles," page 50.

18. a

If you missed 18, go to "Vertical Angles," page 50.

19. d

If you missed 19, go to "Complementary Angles," page 51.

20. b

If you missed 20, go to "Supplementary Angles," page 51.

21. d
22. b
23. d

If you missed 21, 22, or 23, go to "Special Lines and Segments," page 53.

24. a

If you missed 24, go to "Angles Created by Lines and a Transversal," page 55.

25. b

If you missed 25, go to "Angles Created by Lines and a Transversal," page 55.

26. c
27. b

If you missed 26 or 27, go to "Proving Lines Parallel," page 57.

28. c

If you missed 28, go to "Sum of Angle Measure," page 61.

29. c

If you missed 29, go to “Exterior Angles,” in Chapter 3, page 62.

30. a

31. b

If you missed 30 or 31, go to “Classifying Triangles by Angles,” page 65.

32. c

33. b

If you missed 32 or 33, go to “Classifying Triangles by Sides,” page 66.

34. d

If you missed 34, go to “Specially Named Sides and Angles (of a Triangle),” page 68.

35. b

36. c

If you missed 35 or 36, go to “Segments Inside and Outside Triangles,” page 70.

37. b

If you missed 37, go to “Congruent Triangles,” page 73.

38. b

If you missed 38, go to “Proofs of Congruence: SSS and SAS,” page 73.

39. d

If you missed 39, go to “Proofs of Congruence: ASA, SAA, and HL,” page 74.

40. c

If you missed 40, go to “Corresponding Parts (CPCTC),” page 79.

41. b

42. b

If you missed 41 or 42, go to “Isosceles Triangles,” page 81.

43. a

If you missed 43, go to “Triangle Inequality Theorems,” page 82.

44. b

If you missed 44, go to “Polygons,” page 87.

45. d

If you missed 45, go to “Types of Polygons,” page 87.

46. d

If you missed 46, go to "Naming Polygons' Parts," page 88.

47. a

If you missed 47, go to "Number of Sides and Angles (of Polygons)," page 88.

48. b

49. d

If you missed 48 or 49, go to "Angle Sums (of Polygons)," page 89.

50. b

51. d

If you missed 50 or 51, go to "Quadrilaterals," page 92.

52. b

If you missed 52, go to "Trapezoids," page 92.

53. c

If you missed 53, go to "Parallelograms," page 93.

54. d

55. a

If you missed 54 or 55, go to "Proofs of Parallelograms," page 95.

56. c

57. a

58. c

If you missed 56, 57, or 58, go to "Special Parallelograms," page 98.

59. d

If you missed 59, go to "Special Trapezoids," page 101.

60. a

If you missed 60, go to "The Midpoint Theorem," page 102.

61. c

62. b

If you missed 61 or 62, go to "Squares and Rectangles (in Perimeter and Area)," page 105.

63. b

64. a

If you missed 63 or 64, go to "Triangles (in Perimeter and Area)," page 109.

65. c

If you missed 65, go to “Squares and Rectangles (in Perimeter and Area),” page 105.

66. b

67. c

If you missed 66 or 67, go to “Trapezoids (in Perimeter and Area),” page 114.

68. c

69. b

If you missed 68 or 69, go to “Regular Polygons (in Perimeter and Area),” page 116.

70. d

71. c

72. b

If you missed 70, 71, or 72, go to “Circles (in Perimeter and Area),” page 118.

73. c

74. d

If you missed 73 or 74, go to “Ratio (in Similar Figures),” page 123.

75. c

If you missed 75, go to “Proportions (in Similar Figures),” page 124.

76. a

77. b

If you missed 76 or 77, go to “Means and Extremes (in Similar Figures),” page 124.

78. b

79. c

If you missed 78 or 79, go to “Properties of Proportions (in Similar Figures),” page 125.

80. d

If you missed 80, go to “Similar Polygons (in Similar Figures),” page 128.

81. b

82. a

If you missed 81 or 82, go to “Similar Triangles (in Similar Figures),” page 129.

83. c

84. a

If you missed 83 or 84, go to “Proportional Parts of Triangles (in Similar Figures),” page 132.

85. c

If you missed 85, go to "Proportional Parts of Similar Triangles," page 137.

86. d

87. c

If you missed 86 or 87, go to "Perimeter and Areas of Similar Triangles," page 139.

88. b

89. d

If you missed 88 or 89, go to "Geometric Mean (in Right Triangles)," page 143.

90. b

If you missed 90, go to "Altitude to the Hypotenuse (in Right Triangles)," page 144.

91. b

92. d

If you missed 91 or 92, go to "The Pythagorean Theorem (in Right Triangles)," page 147.

93. c

94. a

If you missed 93 or 94, go to "Outgrowths of the Pythagorean Theorem (in Right Triangles)," page 152.

95. a

If you missed 95, go to "Special Right Triangles," page 153.

96. c

97. d

98. b

If you missed 96, 97, or 98, go to "Parts of a Circle," page 159.

99. a

100. d

If you missed 99 or 100, go to "Central Angles and Arcs (in Circles)," page 161.

101. b

If you missed 101, go to "Arcs and Inscribed Angles (in Circles)," page 164.

102. b

103. a

104. d

If you missed 102, 103, or 104, go to "Angles Formed by Chords, Secants, and Tangents (in Circles)," page 167.

105. a

106. c

If you missed 105 or 106, go to "Segments of Chords, Secants, and Tangents (in Circles)," page 174.

107. b

108. c

If you missed 107 or 108, go to "Arc Lengths and Sectors (in Circles)," page 177.

109. b

110. c

111. d

If you missed 109, 110, or 111, go to "Prisms (in Solid Geometry)," page 183.

112. c

113. d

If you missed 112 or 113, go to "Right Circular Cylinders (in Solid Geometry)," page 189.

114. d

115. c

116. b

If you missed 114, 115, or 116, go to "Pyramids (in Solid Geometry)," page 193.

117. c

118. c

119. b

If you missed 117, 118, or 119, go to "Right Circular Cones (in Solid Geometry)," page 195.

120. b

121. c

If you missed 120 or 121, go to "Spheres (in Solid Geometry)," page 197.

122. b

123. c

124. d

If you missed 122, 123, or 124, go to "Locating Points on Coordinate Axes," page 201.

125. d

If you missed 125, go to "The Distance Formula," page 203.

126. a

If you missed 126, go to "The Midpoint Formula," page 205.

127. b

128. a

If you missed 127 or 128, go to "Slope of a Line," page 208.

129. c

130. a

If you missed 129 or 130, go to "Slope of Parallel and Perpendicular Lines," page 210.

131. a

132. b

If you missed 131 or 132, go to "Equations of Lines," page 211.

Chapter 1

Basic Geometric Ideas

The word *geometry* comes from two ancient Greek words, *ge*, meaning earth, and *metria*, meaning measure. So, literally, geometry means to measure the earth. It was the first branch of math that began with certain assumptions and used them to draw more complicated conclusions. Over time, geometry has become a body of knowledge that helps us to logically create chains of conclusions that let us go from knowing certain things about a figure to predicting other things about it with certainty. Although a little arithmetic and a little algebra are used in building an understanding of geometry, this branch of math really can stand on its own, as a way of constructing techniques and insights that may help you to better understand later mathematical ideas, and that, believe it or not, may help you to live a more fulfilling life.

Naming Basic Forms

The bulk of this book deals with plane geometry—that is, geometry on a perfectly flat surface. Many different types of plane figures exist, but all of them are made up of a few basic parts. The most elementary of those parts are points, lines, and planes.

Points

A **point** is the simplest and yet most important building block in geometry. It is a location and occupies no space. Because a point has no height, length, or width, we can't actually draw one. This is true of many geometric parts. We can, however, *represent* a point, and we use a dot to do that. We name points with single uppercase letters.

This diagram shows three dots that represent points C, M, and Q.

Lines

Lines are infinite series of points. *Infinite* means without end. A line extends infinitely in two opposite directions, but has no width and no height. Just to be clear, in geometry, *line* and *straight line* mean the same thing. Contrary to the popular notion, a line is *not* the shortest distance between two points. (We'll come back to this later.)

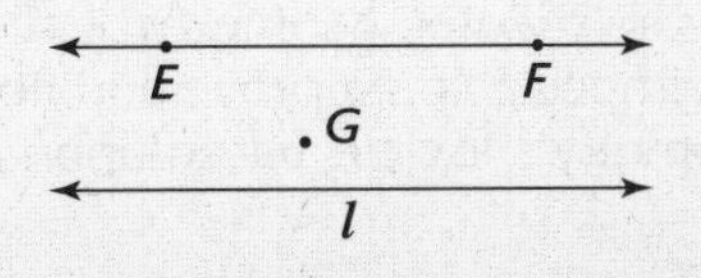

A line may be named by any two points on it, as is line EF, represented by the symbol $\overleftrightarrow{EF}$ or $\overleftrightarrow{FE}$. It may also be named by a single lowercase letter, as is line l.

Points that are on the same line are said to be **collinear points**. Point *E* and Point *F* in the preceding diagram are collinear points. Point *G* is not collinear with *E* and *F*. Taken altogether, it may be said that *E*, *F*, and *G* are **noncollinear points**. You'll see why this distinction is important a little later in this chapter.

Planes

A **plane** is an infinite set of points extending in all directions along a perfectly flat surface. It is infinitely long and infinitely wide. A plane has a thickness (or height) of zero.

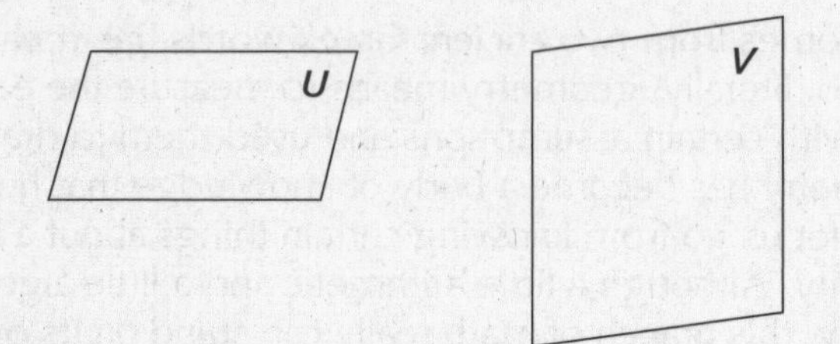

A plane is named by a single uppercase letter and is often represented as a four-sided figure, as in planes *U* and *V* in the preceding diagram.

Example Problems

These problems show the answers and solutions.

1. What is the maximum number of lines in a plane that can contain two of the points *A*, *B*, and *C*?

   ***Answer:* 3** Consider that two points name a line. It is possible to make three sets of two points from the three letters: *AB*, *AC*, and *BC*. That means it's possible to form three unique lines: $\overleftrightarrow{AB}$, $\overleftrightarrow{AC}$, and $\overleftrightarrow{BC}$. See the following figure.

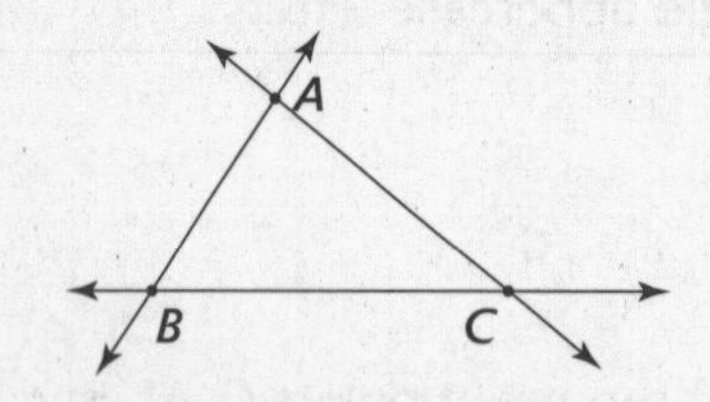

2. What kind of geometric form is the one named *H*?

   ***Answer:* not enough information** A single uppercase *H* could be used to designate a point or a plane.

Postulates and Theorems

As noted at the very beginning of the chapter, geometry begins with assumptions about certain things that are very difficult, if not impossible, to prove and flows on to things that can be proven. The assumptions that geometry's logic is based upon are called **postulates**. Sometimes, you may see them referred to as **axioms**. The two words mean essentially the same thing, Here are the first six of them, numbered so that we can refer back to them easily:

Postulate 1: A line contains at least two points.
Postulate 2: A plane contains a minimum of three noncollinear points.
Postulate 3: Through any two points there can be exactly one line.
Postulate 4: Through any three noncollinear points there can be exactly one plane.
Postulate 5: If two points lie in a plane, then the line they lie on is in the same plane.
Postulate 6: Where two planes intersect, their intersection is a line.

From these six postulates it is possible to prove these **theorems**, numbered for the same reason:

Theorem 1: If two lines intersect, they intersect in exactly one point.
Theorem 2: If a point lies outside a line, then exactly one plane contains the line and the point.
Theorem 3: If two lines intersect, then exactly one plane contains both lines.

Example Problems

These problems show the answers and solutions. State the postulate or theorem that may be used to support the statement made about each diagram.

1. There is another point on line l in addition to R.

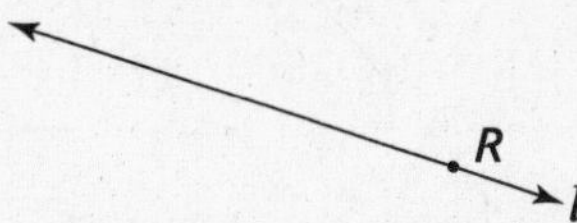

Answer: **A line contains at least two points.** (Postulate 1)

2. Only one line contains point M and point N.

N
M

Answer: **Through any two points there can be exactly one line.** (Postulate 3)

Work Problems

Use these problems to give yourself additional practice. State the postulate or theorem that may be used to support the statement made about each diagram.

1. Lines m and l are in the same plane.

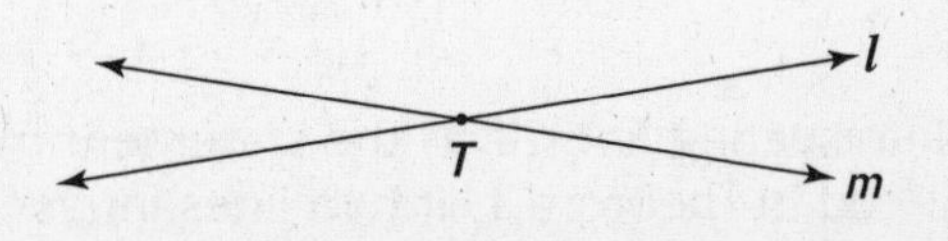

2. There is no other intersection for *n* and *p* other than B.

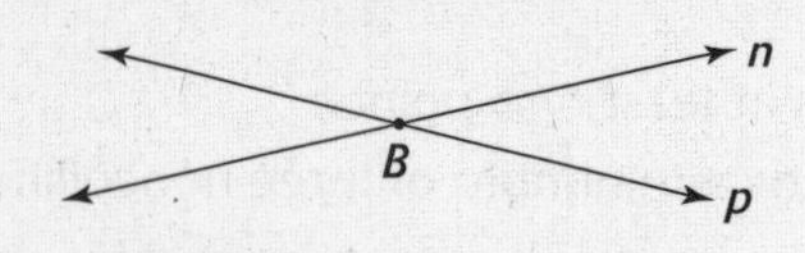

3. Point *F* and $\overleftrightarrow{DE}$ are in the same plane.

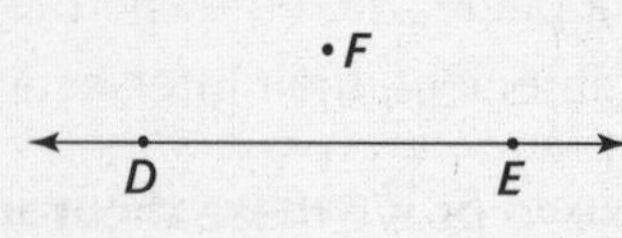

4. Points *J*, *K*, and *L* are all in the same plane.

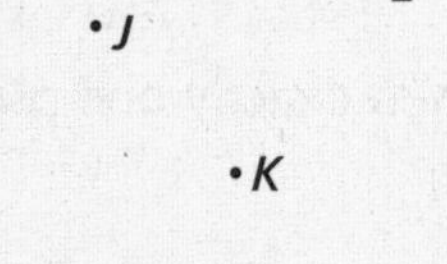

5. The intersection of planes *P* and *Q* is line *r*.

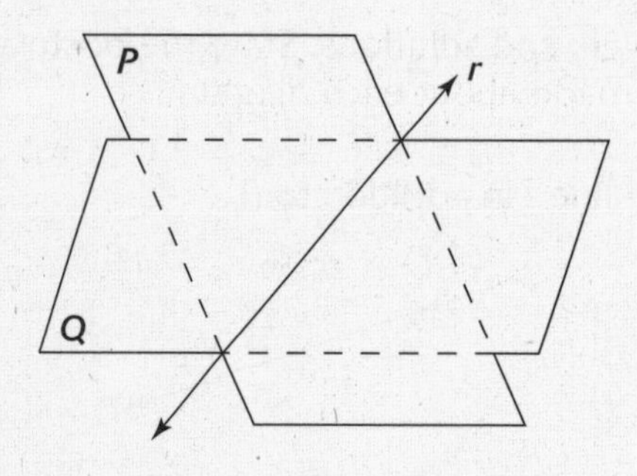

6. $\overleftrightarrow{AB}$ lies in plane *W*.

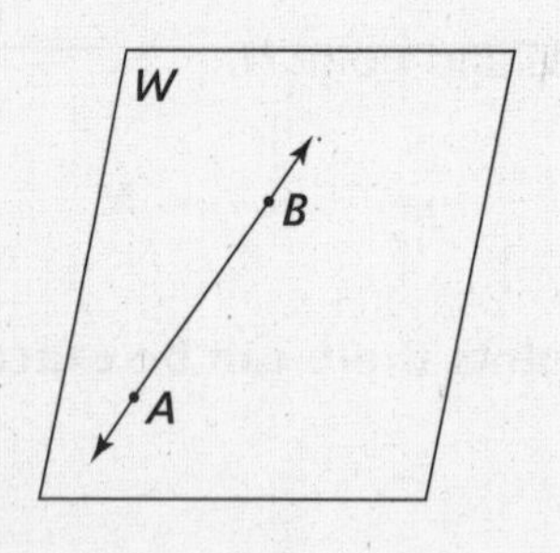

Worked Solutions

1. The figure shows two intersecting lines, and the statement mentions a plane. That relationship is dealt with by Theorem 3: If two lines intersect, then exactly one plane contains both lines.

2. The figure shows two intersecting lines, and the statement mentions the point of intersection. That's covered in Theorem 1: If two lines intersect, they intersect in exactly one point.

3. This figure concerns a line and a noncollinear point, and the statement mentions a plane. That's Theorem 2: If a point lies outside a line, then exactly one plane contains the line and the point.

4. We are shown three noncollinear points, and a plane is mentioned. That's Postulate 4: Through any three noncollinear points there can be exactly one plane.

5. Here, we have two intersecting planes and line *r*. That's Postulate 6: Where two planes intersect, their intersection is a line.

6. The diagram shows a line in a plane, but two points on that line are clearly marked. That should lead us straight to Postulate 5: If two points lie in a plane, then the line they lie on is in the same plane.

Finding Segments, Midpoints, and Rays

You'll recall that geometry means earth measure. We've already dealt with the concept of lines, but because lines are infinite, they can't be measured. Much of geometry deals with parts of lines. Some of those parts are very special—so much so that they have their own special names and symbols. The first such part is the **line segment**.

Line Segments

A line segment is a finite portion of a line and is named for its two endpoints.

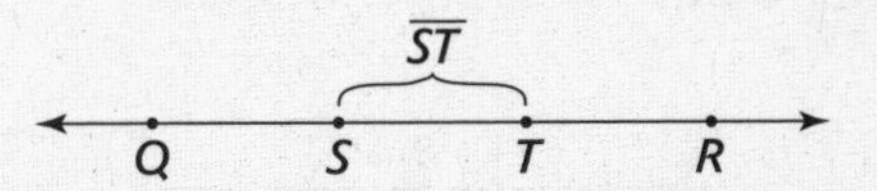

In the preceding diagram is segment $\overline{ST}$. Notice the bar above the segment's name. Technically, $\overline{ST}$ refers to points *S* and *T* and all the points in between them. *ST*, without the bar, refers to the distance from *S* to *T*. You'll notice that $\overline{ST}$ is a portion of $\overleftrightarrow{QR}$.

Each point on a line or a segment can be paired with a single real number, which is known as that point's **coordinate**. The distance between two points is the absolute value of the difference of their coordinates.

If $b > a$, then $AB = b - a$. This postulate, number 7, is known as the *Ruler Postulate*.

Example Problems

These problems show the answers and solutions.

E F G H I J K

3 5 7 9 11 13 15

1. Find the length of $\overline{EH}$, or, put more simply, find *EH*.

   ***Answer:* 6** To find the length of $\overline{EH}$, first find the coordinates of point *E* and point *H*.

E's coordinate is 3, and point H's coordinate is 9.

$$EH = 9 - 3$$
$$EH = 6$$

2. Find the length of $\overline{FK}$, or, put more simply, find FK.

***Answer:* 10** To find FK, first find the coordinates of point F and point K.

F's coordinate is 5, and point K's coordinate is 15.

$$FK = 15 - 5$$
$$FK = 10$$

Segment Addition and Midpoint

Postulate 8 is known as the Segment Addition Postulate. It goes like this:

> **Postulate 8 (Segment Addition Postulate):** If N lies between M and P on a line, then $MN + NP = MP$. This is, in fact, one of many postulates and theorems that can be restated in general terms as the whole is the sum of its parts.

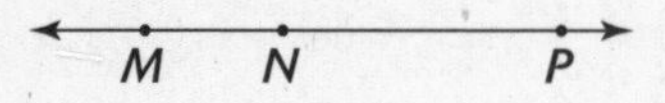

The midpoint of a line segment is the point that's an equal distance from both endpoints. B is the midpoint of $\overline{AC}$ because $\overline{AB} = \overline{BC}$.

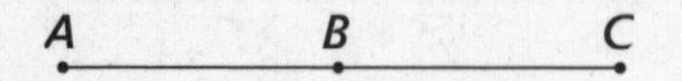

This brings us to the obvious fact stated by Theorem 4:

> **Theorem 4:** A segment has exactly one midpoint.

Example Problems

These problems show the answers and solutions.

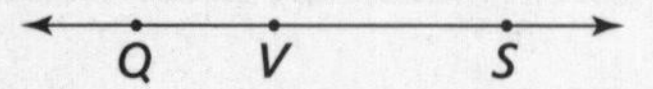

1. V lies between Q and S. Find QS if $QV = 6$ and $VS = 10$.

***Answer:* 16** Since V lies between Q and S, Postulate 8 tells us that

$$QV + VS = QS$$
$$6 + 10 = 16$$
$$QS = 16$$

2. Find the midpoint of $\overline{RZ}$.

R S T U V W X Y Z
7 10 13 14 19 21 23 25 31

***Answer:* V** We can solve this in two ways. First consider the coordinates of the endpoints, *R* (7) and *Z* (31). Their difference is 24 ($31 - 7 = 24$). That means that the segment is 24 units long, so its midpoint must be half of 24, or 12 units from either endpoint. Adding 12 to the 7 (*R*'s coordinate) gives us 19, the coordinate of *V*. *V* is the midpoint.

The other method is to take the average of the coordinates of the two endpoints, which is done by adding them together and dividing by 2.

$$\frac{7+31}{2} = \frac{38}{2}$$

$$\frac{38}{2} = 19$$

And 19, of course, is the coordinate of midpoint, *V*.

Rays

Geometric rays are like the sun's rays. They have a beginning point (or endpoint), and they go on and on without end in a single direction. The sun's rays don't have names (as far as we know). A geometric ray is named by its endpoint and any other point on it.

Above is $\overrightarrow{DE}$ (read ray *DE*). The arrow above the letters not only indicates that the figure is a ray. It also indicates the direction in which the ray is pointing. The arrow's head is over the non-endpoint.

G F

This is ray *FG*. It may be written $\overrightarrow{FG}$, or $\overleftarrow{GF}$.

Work Problems

Use these problems to give yourself additional practice.

M N

1. Is the preceding figure more accurately named $\overline{MN}$ or $\overline{NM}$?

P R Q

2. *PR* is 8, and *PQ* is 22. What is the length of $\overline{RQ}$?

A (6), B (12), C (18), D (26), E (30), F (36), G (40), H (46), I (48), J (52), K (54)

3. Find the midpoint of $\overline{AE}$.

4. Find the midpoint of $\overline{AK}$.

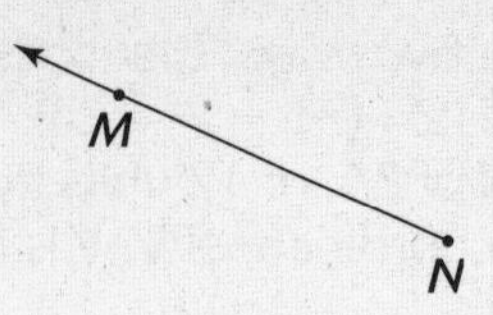

5. Name the ray here.

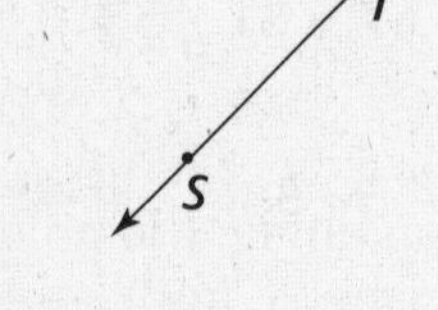

6. Name the ray here.

Worked Solutions

1. **Both names are suitable.** A segment does not have direction, so the order in which its endpoints are named is not relevant.

2. **14** By the Segment Addition Postulate,

$$PR + RQ = PQ$$
$$\text{Then } 8 + RQ = 22$$
$$\text{So } RQ = 22 - 8$$
$$RQ = 14$$

3. **C** $\overline{AE}$'s endpoints are A and E with coordinates of 6 and 30. That means the length

$$AE = 30 - 6$$
$$AE = 24$$

The midpoint is half of 24, or 12 from either endpoint. 12 from the starting coordinate of 6 is 18, the coordinate of point C.

4. **E** This is essentially the same problem as 3, but we'll use the alternate method to solve it. Let's average the endpoints' coordinates.

First, add them together: $54 + 6 = 60$

Then, divide by 2: $\frac{60}{2} = 30$

30 is the coordinate of point E.

5. $\overrightarrow{NM}$ **or** $\overleftarrow{MN}$ The only thing that matters in the naming of a ray is which end has the endpoint and which point is elsewhere on the ray. The endpoint will be under the blunt end of the arrow, and the point that is elsewhere on the ray goes under the arrowhead.

6. $\overrightarrow{TS}$ **or** $\overleftarrow{ST}$ See the explanation for problem 5.

Angles and Angle Pairs

Angles are as important as line segments when it comes to forming geometric figures. Without them, there would be no plane figures, with the possible exception of circles.

Forming and Naming Angles

An angle is formed by two rays that have a common (or shared) endpoint. The rays form the **sides** of the angle, and their endpoint forms its **vertex**. The measurement of the opening of an angle is expressed in degrees. The smallest angle practical has a degree measure of 0°. Imagine the angle formed by the hands of an analog clock at noon. That is a 0° angle. There is no real limit to the upper number of degrees in an angle, but no unique angle contains more than 359°, since an angle of 360° is indistinguishable from one of 0°.

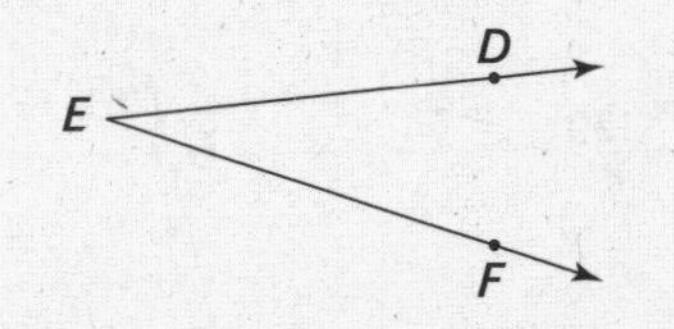

Rays $\overrightarrow{ED}$ and $\overrightarrow{EF}$ form the sides of this angle, shown by the symbol $\angle DEF$, whose vertex is at E.

Here we have two pictures of the same angle.

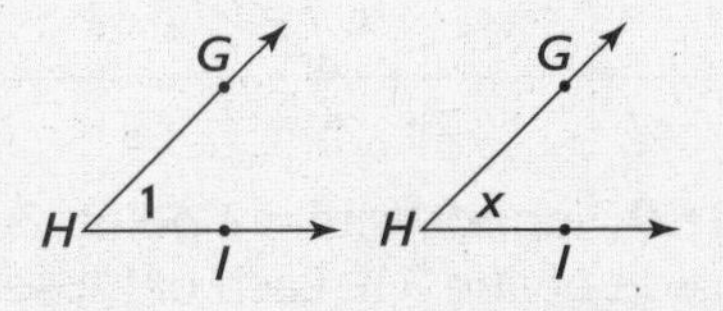

We need two pictures in order to cover all the different ways there are to name this angle. It may be named by three letters—one from each ray and the vertex, with the name of the vertex always being in the middle. Thus, it is $\angle GHI$ or $\angle IHG$. If there is no chance of ambiguity (more than one angle at the vertex), it may be named by the vertex alone, hence $\angle H$. It may also be named by a number or a lowercase letter inside the angle, so it is also $\angle 1$, on the left, or $\angle x$, on the right.

Example Problems

These problems show the answers and solutions.

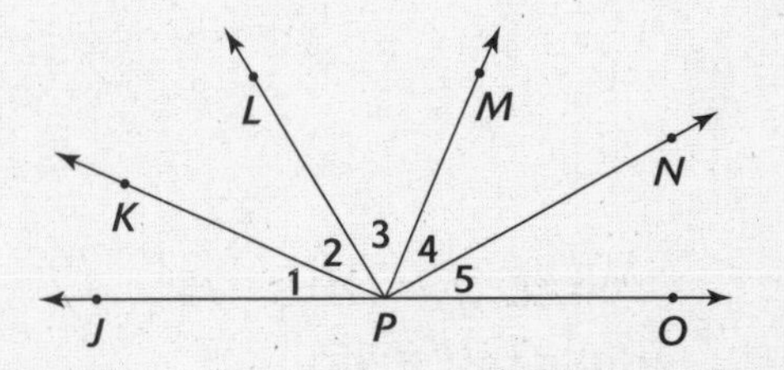

1. State another name for $\angle 3$.

   Answer: $\angle LPM$ **or** $\angle MPL$ Use the name of one point on each ray with the designation of the vertex in the middle. $\angle P$ would not do, since there are 5 separate angles at vertex P.

2. What is another name for $\angle KPL$?

Answer: $\angle LPK$ or $\angle 2$ The first is the reverse of the order of the letters in the question (always legitimate), or second, the number that is written in the opening of that angle.

3. What is another name for the sum of angles 4 and 5?

Answer: $\angle OPM$ or $\angle MPO$ This wasn't really a fair question, since we haven't yet discussed angle addition, still it is a natural application of the already noted fact that a whole is the sum of its parts. The two angles combined into one are bounded by $\overrightarrow{PM}$ and $\overrightarrow{PO}$, hence the two choices given.

The Protractor Postulate and Addition of Angles

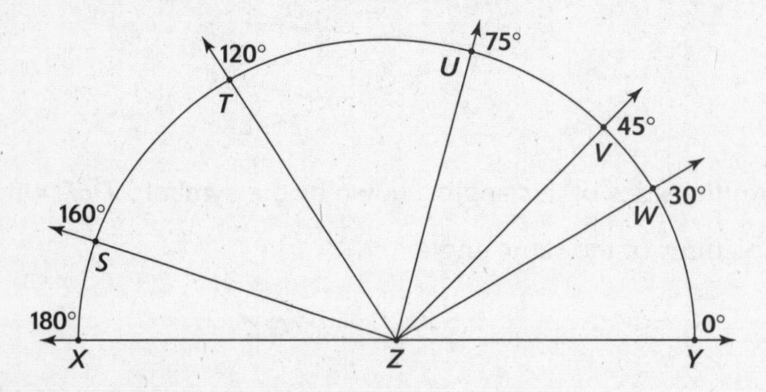

The Protractor Postulate, Postulate 9, supposes that a point, Z, exists on line XY. Think of all rays with endpoint Z that exist on one side of line XY. Each of those rays may be paired with exactly one number between 0° and 180°, as you can see in the preceding figure. The positive difference between two numbers representing two different rays is the degree measure of the angle with those rays as its sides. So, the measure of $\angle VZW$ (represented $m\angle VZW$) = 45° − 30° = 15°.

Postulate 10 is the Addition of Angles postulate. Simply put, if $\overrightarrow{ZS}$ lies between $\overrightarrow{ZX}$ and $\overrightarrow{ZT}$, then $m\angle XZT = m\angle XZS + m\angle SZT$. It's just another statement of the whole equaling the sum of its parts, as already used in the previous problems.

Example Problems

These problems show the answers and solutions. Use the preceding diagram to solve these problems.

1. Find $m\angle TZU$.

Answer: 45°

$$m\angle TZU = 120° - 75°$$
$$m\angle TZU = 45°$$

2. Find $m\angle UZY$.

Answer: 75°

$$m\angle UZY = 75° - 0°$$
$$m\angle UZY = 75°$$

3. Find $m\angle SZV$.

Answer: **115°**

$$m\angle SZV = 160° - 45°$$
$$m\angle SZV = 115°$$

Angle Bisector

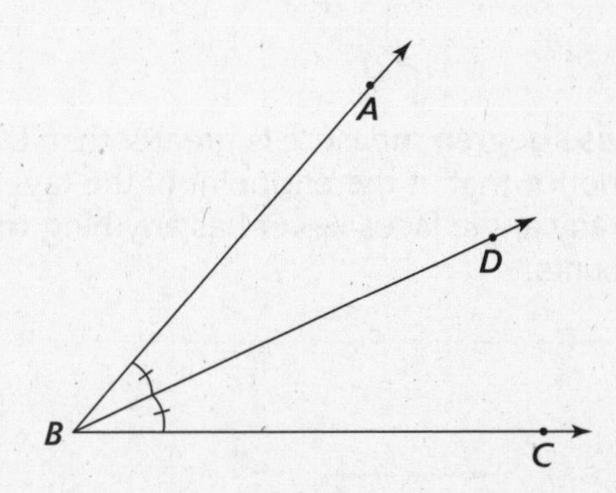

An **angle bisector** is a ray that divides an angle into two angles of equal degree measure. In the figure immediately preceding, $m\angle ABD$ is marked as being equal to $m\angle DBC$, therefore $\overrightarrow{BD}$ is the angle bisector of $\angle ABC$.

Certain angles have special names, and we'll begin looking at them next, but first, consider the following:

Theorem 5: An angle that is not a straight angle has only one angle bisector.

Think about why that is.

Right Angles

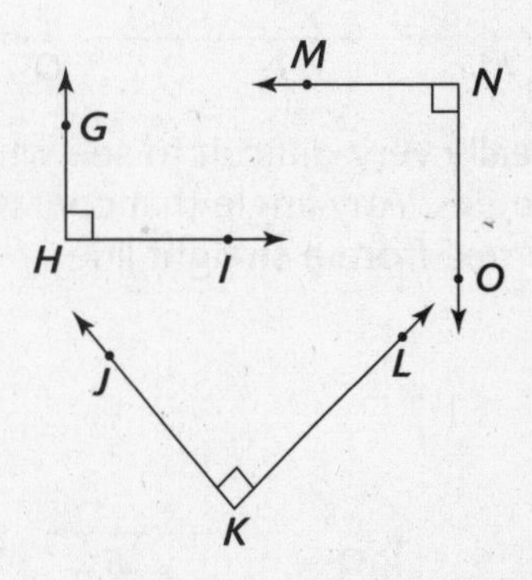

A **right angle** is an angle whose measure is 90°. Angles *GHI*, *JKL*, and *MNO* are all right angles. That little corner in each of those right angles should serve as a reminder that they have the same shape as the corners of a book. That brings us to the following theorem:

Theorem 6: All right angles are equal.

Acute Angles

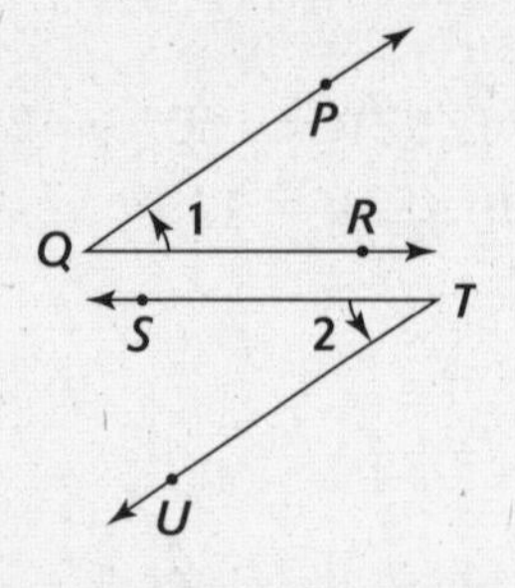

An **acute angle** is an angle whose degree measure is greater than 0° and less than 90°. The word *acute* means sharp, and you'll notice that at the endpoint of the rays that form $\angle 1$ and $\angle 2$ is a sharp point. Note that the way any angle faces never has anything to do with the kind of angle it is. Only the degree measure counts.

Obtuse Angles

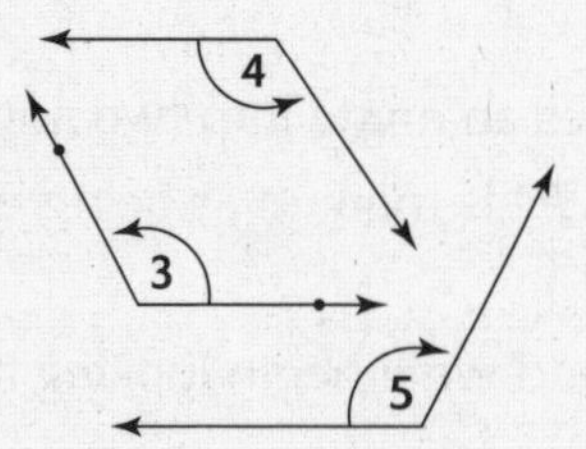

An **obtuse angle** has a degree measure greater than 90° and less than 180°. Angles 3, 4, and 5 are three examples of obtuse angles. *Obtuse* means dull or blunt. Compare the shapes to that of a sharp acute angle.

Straight Angles

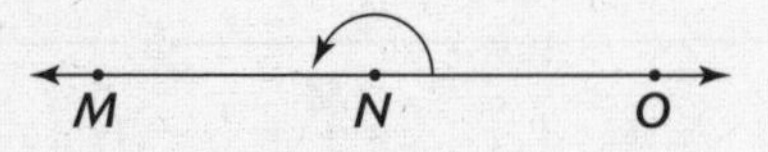

$\angle MNO$ is a **straight angle**. It's not really very difficult to see why. The degree measure of a straight angle is exactly 180°, or two right angles. Any angle that contains exactly 180° is a straight angle and indistinguishable, save for its vertex, from a straight line.

Reflex Angles

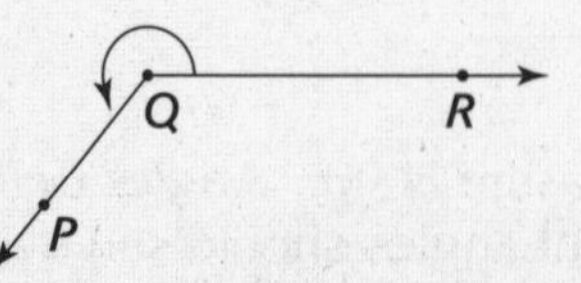

A **reflex angle** is formed by a pair of rays where one of them has remained stationary and the other one has rotated through an angle of greater than 180° and less than 360°. You are not likely to encounter many reflex angles in your study of geometry, but you should be aware of their existence.

Example Problems

These problems show the answers and solutions.

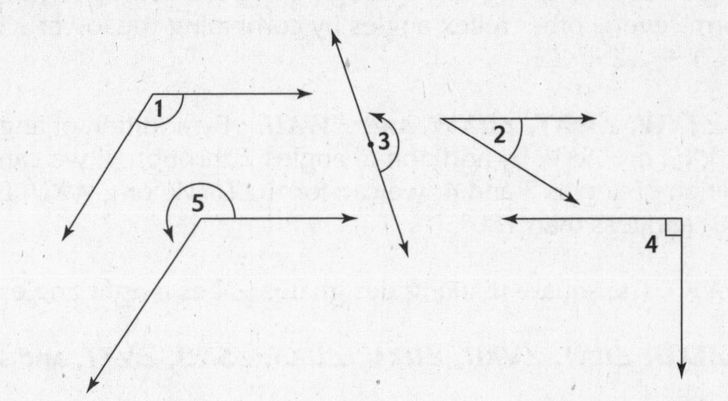

1. Which of the angles shown is a right angle?

   ***Answer:* ∠4** The square inside the angle tells us that ∠4 is a right, or 90° angle.

2. Which of the angles shown is an acute angle?

   ***Answer:* ∠2** An acute angle must measure fewer than 90°. The fact that ∠2 comes to an acute (sharp) point is the tip-off.

3. Which of the angles shown is an obtuse angle?

   ***Answer:* ∠1** The measures of three of the angles are greater than 90°, but only ∠1's measure is both greater than 90° and smaller than 180°—the definition of an obtuse angle.

Work Problems

Use these problems to give yourself additional practice. All of the work problems in this section refer to this figure.

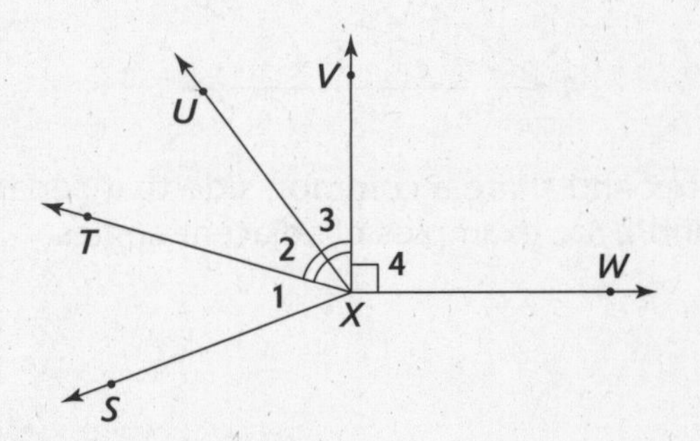

1. Name a reflex angle shown in the diagram.
2. Name two obtuse angles shown in the figure.
3. Give the letter name of a right angle in the diagram.
4. Name as many acute angles as you can find in the diagram, using letter designations.
5. Name the angle bisector.

Worked Solutions

1. **∠*WXS*** By angle addition of angles 1 through 4, we are able to form reflex angle *WXS*. We can also form several other reflex angles by combining the lower ∠*WXS* with ∠ 1, ∠1 +∠2, and ∠1 + ∠2 + ∠3.

2. **∠*VXS*, ∠*SXV*, ∠*TXW*, ∠*WXT*, ∠*UXW*, and ∠*WXU*** By addition of angles 1 through 3, we can form ∠*VXS*, or ∠*SXV*; by addition of angles 2 through 4, we can form ∠*TXW*, or ∠*WXT*; by addition of angles 3 and 4, we can form ∠*UXW*, or ∠*WXU*. Each of those is greater than 90° and less than 180°.

3. **∠*VXW* or ∠*WXV*** The square marking designates ∠4 as a right angle.

4. **∠*TXS*, ∠*SXT*, ∠*TXU*, ∠*UXT*, ∠*VXU*, ∠*UXV*, ∠*UXS*, ∠*SXU*, ∠*VXT*, and ∠*TXV***

   ∠1 is ∠*TXS*, or ∠*SXT*, ∠2 is ∠*TXU*, or ∠*UXT*, and ∠3 is ∠*VXU*, or ∠*UXV*.

   ∠1 + ∠2 make ∠*UXS*, ∠*SXU*; ∠2 + ∠3 make ∠*VXT*, and ∠*TXV*. All are clearly less than 90°.

5. $\overrightarrow{XU}$ Angles 2 and 3 are marked as being of equal measure. An angle bisector is defined as a line that separates an angle into two angles of equal measure.

Special Angle Pairs

Certain pairs of angles have special names. Those names are in some cases based upon their positions relative to one another. In other cases they are based upon the angles' degree measures adding up to a certain amount.

Adjacent Angles

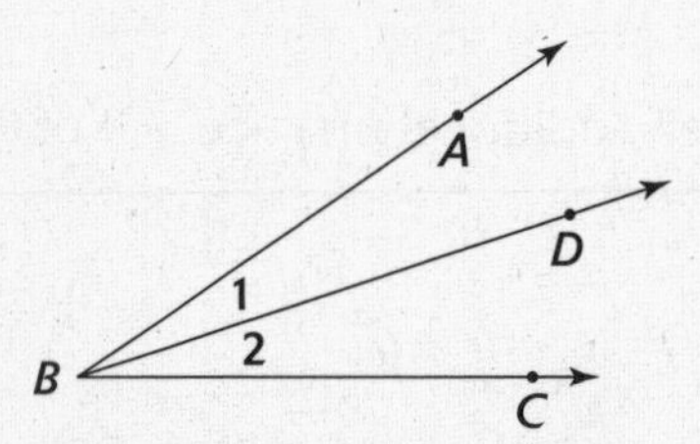

Two angles that share a vertex and share a common side that separates them are known as **adjacent angles**. Angles 1 and 2 are examples of adjacent angles.

Vertical Angles

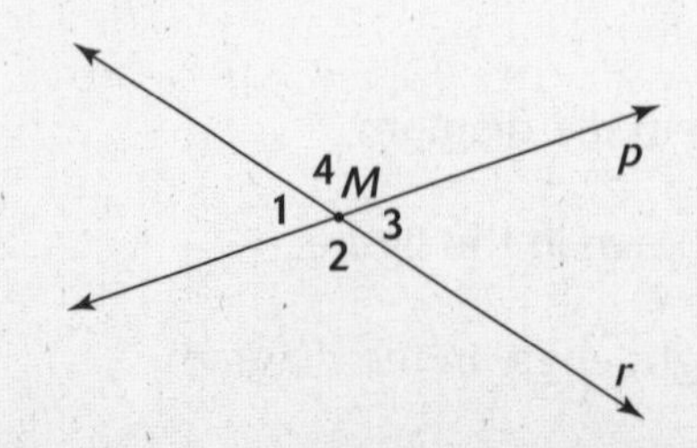

When two lines intersect so as to form four angles, the angles on opposite sides of the common vertex are known as **vertical angles**. Of the four angles formed at *M*, ∠1 and ∠3 are vertical angles. So are ∠2 and ∠4.

Theorem 7 tells us that vertical angles are equal in measure. That means $m\angle 1 = m\angle 3$, and $m\angle 2 = m\angle 4$.

This pair of lines also forms four pairs of adjacent angles: $\angle 1$ and $\angle 2$, $\angle 2$ and $\angle 3$, $\angle 3$ and $\angle 4$, $\angle 4$ and $\angle 1$.

Complementary Angles

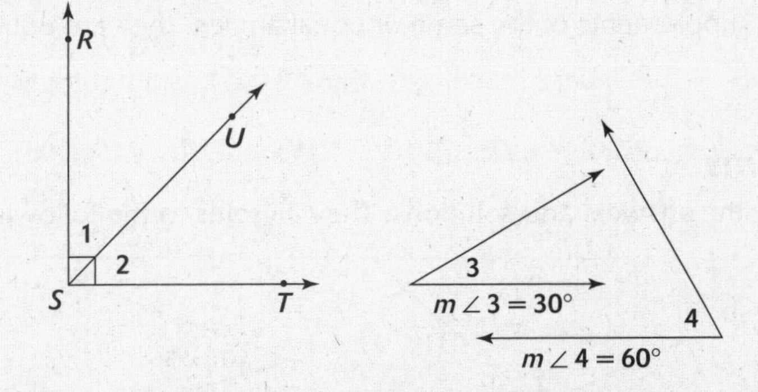

Any two angles that add up to 90° are called **complementary angles**. $\angle 1$ and $\angle 2$ are adjacent complementary angles. $\angle 1$ is said to be the **complement** of $\angle 2$. $\angle 2$ is said to be the complement of $\angle 1$. Notice that the word has no "i" in it. It's complement, not compliment. Angles 3 and 4 are **nonadjacent** complementary angles.

That brings us to **Theorem 8**, which states: If two angles are complements of the same or equal angles, they are equal to each other.

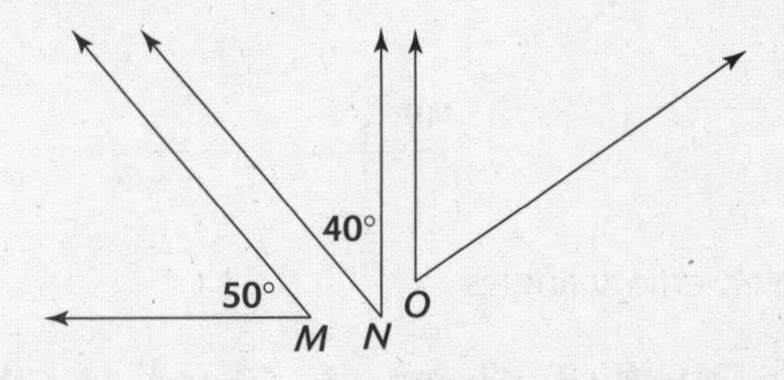

$m\angle M = 50°$. $\angle M$ is complementary to $\angle N$. $\angle O$ is also complementary to $\angle N$. That means that $m\angle O$ must be 50°, since both $\angle M$ and $\angle O$ are complements of the same angle.

Supplementary Angles

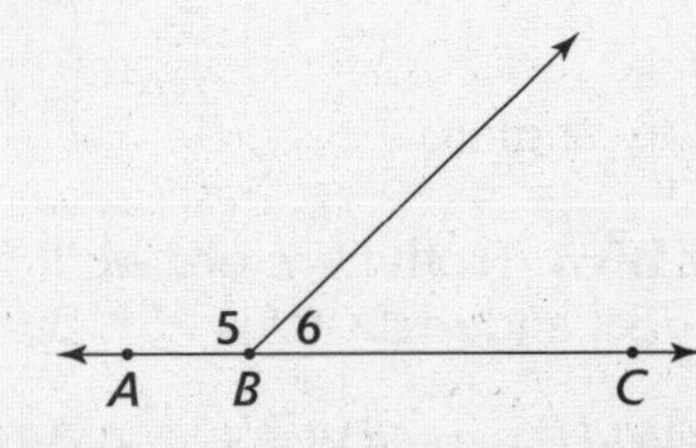

When two angles add up to a total of 180°, they're called **supplementary angles**. $\angle 5$ and $\angle 6$ are adjacent supplementary angles. That gives us:

Theorem 9: If two adjacent angles have their noncommon sides lying on a line, then they are supplementary angles. That's pretty much a no-brainer.

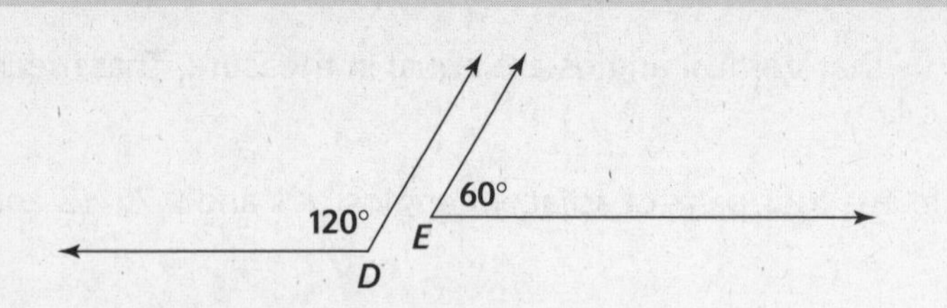

∠*D* and ∠*E* are nonadjacent supplementary angles.

Theorem 10, does for supplementary angles what Theorem 8 did for complementary ones: If two angles are supplements of the same or equal angles, they are equal to each other.

Example Problems

These problems show the answers and solutions. They all refer to the following diagram.

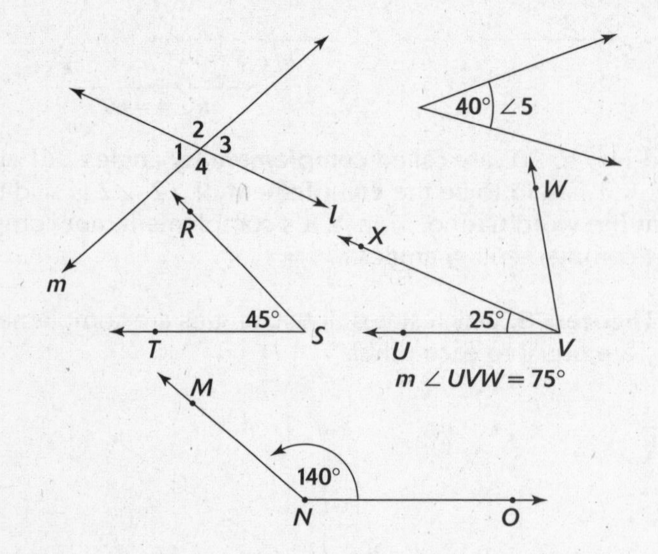

1. Identify a pair of supplementary angles.

   ***Answer:* ∠1 and ∠2, ∠2 and ∠3, ∠3 and ∠4, ∠4 and ∠1** When two lines intersect, each pair of adjacent angles is also a pair of supplementary angles.

2. Identify a pair of vertical angles.

   ***Answer:* ∠1 and ∠3, ∠2 and ∠4** Vertical angles are nonadjacent angles formed by a pair of intersecting lines.

3. Identify a pair of complementary angles.

   Answer:* ∠5 and ∠*XVW* or ∠*WVX (From here on, we'll name each angle only once.) ∠*WVU* has angle measure of 75°, and *m*∠*XVU* is 25°. By the Angle Addition Postulate,

   $$m\angle WVU = m\angle XVW + m\angle XVU$$

   That means $75° = m\angle XVW + 25°$

$$\text{So } 50° = m\angle XVW$$
$$\text{Or } m\angle XVW = 50°.$$

Complementary angles total 90°, so the complement of a 50° $\angle XVW$ is the 40° $\angle 5$.

Work Problems

Use these problems to give yourself additional practice

1. Can an angle be its own complement? Give an example.
2. Can an angle be its own supplement? Give an example.
3. Is it possible to have more than two pairs of vertical angles at a given point? Explain.
4. Can two adjacent angles formed by intersecting lines at a single point not be supplementary? Explain.
5. If two angles are supplementary and one of them is acute, what must be true of the second angle?

Worked Solutions

1. **Yes** Since there are 90° in a right angle, half of 90, or 45° is its own complement.
2. **Yes** Since there are 180° in a straight angle, half of 180, or 90° is its own supplement.
3. **Yes** If 3 lines intersect, there'll be three pairs, 4 lines, 4 pairs, and so on. Actually you can have more than that if you double up the angles or even triple them up, but why bother?
4. **Yes** The adjacent angles are supplementary if and only if exactly two lines cross. If three or more lines cross, the measures of any two adjacent angles would sum to less than 180°.
5. **It must be obtuse.** Since an acute angle's measure is less than 90°, its supplement's measure must be more than 90° for them to sum to 180°.

Special Lines and Segments

You've probably stood in line at an ice cream parlor or waiting to get in to a concert. If you live in a city, lines are a part of everyday life; in a small town, not so much so. In geometry, there are three types of lines that are of interest, and it is essential for students to understand each of them.

Intersecting Lines and Segments

Two or more lines that meet or cross at a point are called **intersecting lines**. The point at which they meet is considered to be a part of both lines. Here we see line j intersecting line k at M.

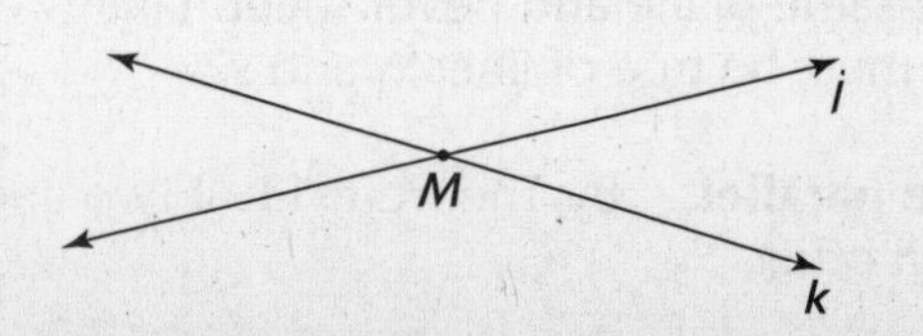

Perpendicular Lines and Segments

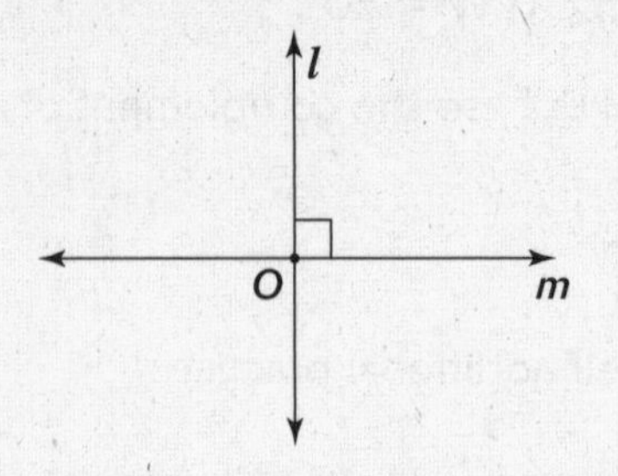

Two intersecting lines that form right angles are known as perpendicular lines. They are denoted by the symbol, $\perp$. In the preceding figure, line $l \perp$ line m.

Parallel Lines and Segments

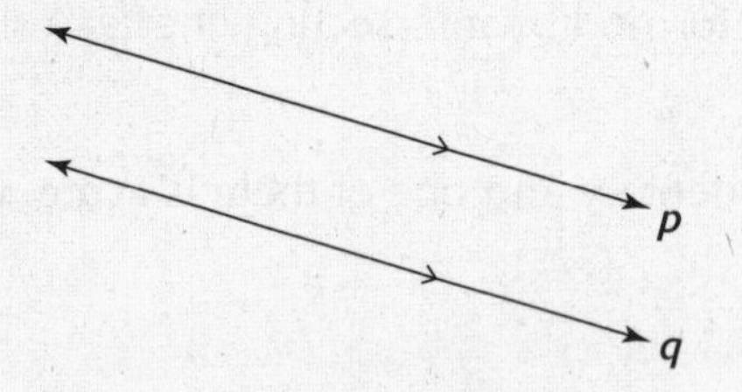

Two lines in the same plane that never intersect are called parallel lines. We'll discuss them in detail in the next chapter. The symbol || means is parallel to and is used as a shorthand for parallel lines. So, in this diagram, line p || line q. Notice that arrow heads mark the lines as being parallel.

Theorem 11: If two lines are parallel to a third line, they are parallel to each other.

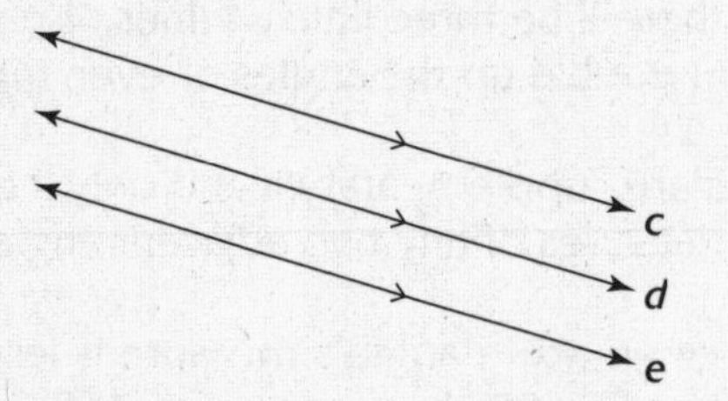

Line c || line d. Line e || line d. Therefore, by Theorem 11, line c || line e.

Example Problems

These problems show the answers and solutions.

1. Lines v and w intersect at a non-right angle. What kind of lines are they?

   ***Answer:* Intersecting lines** Parallel lines never cross, and perpendicular lines cross at a right angle. Only intersecting lines remain.

2. Lines q and r are in the same plane and never meet. Lines r and s are in the same plane and never meet. What must be true of lines q and s?

   ***Answer:* They must be parallel.** By Theorem 11, if two lines are parallel to a third line, they are parallel to each other.

Chapter 2

Parallel Lines

The real world presents us with many examples of parallel and intersecting line segments. Railroad tracks are parallel, for example, and streets may be intersecting or parallel. Parallel lines are really of use only for the purposes of geometry, since in reality, no physical objects are infinite. Still, by studying parallel lines, we can learn a great deal about real-world relationships.

Angles Created by Lines and a Transversal

A line that crosses two or more other lines in the same plane is known as a **transversal** to those lines. In the following figure, *t* is a transversal to *l* and *m*.

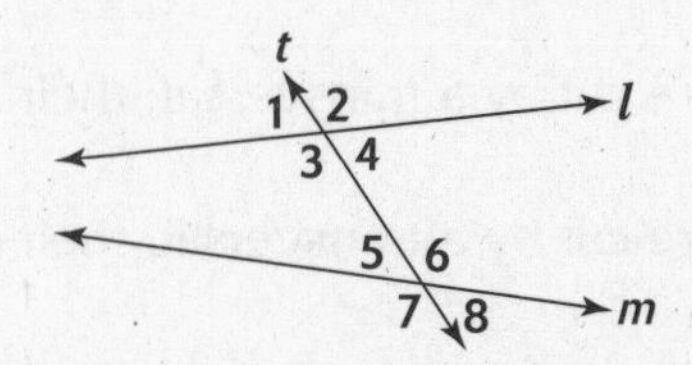

As you can see, the two lines and the transversal form eight angles at the two points of intersection. Certain pairs of those angles with different vertices are given special names. Those in the same relative positions on each of the lines such as above each line and left of the transversal or below each line and left of the transversal, or above and right or below and right are called **corresponding angles**. $\angle 2$ and $\angle 6$ are corresponding angles.

Angles between the lines and on opposite sides of the transversal are known as **alternate interior angles**. $\angle 3$ and $\angle 6$ are alternate interior angles, as are $\angle 4$ and $\angle 5$. *Tip:* To recognize alternate interior angles, think of the letter, "Z," either forward or reversed. Also formed are **alternate exterior angles** $\angle 1$ and $\angle 8$ being a pair of those, **consecutive interior angles** (same side interior) such as $\angle 3$ and $\angle 5$, and **consecutive exterior angles**, like $\angle 2$ and $\angle 8$. Of all of those, the most important for our purposes are the first two types.

Example Problems

These problems show the answers and solutions.

1. Name another pair of corresponding angles that are on the same side of transversal *t* as $\angle 2$ and $\angle 6$.

   ***Answer:* $\angle 4$ and $\angle 8$** $\angle 2$ and $\angle 6$ are on the right side of the transversal *t* and above the lines. $\angle 4$ and $\angle 8$ are on the right side of the transversal *t* and below the lines.

2. Name all the remaining corresponding angles on the diagram.

***Answer:* $\angle 1$ and $\angle 5$; $\angle 3$ and $\angle 7$** All corresponding angles to the right of *t* were found in question 1. That leaves the two pairs remaining to the left of *t*. Those are $\angle 1$ and $\angle 5$ above the lines and $\angle 3$ and $\angle 7$ below the lines.

Angles Created by Parallel Lines and a Transversal

The Parallel Postulate (Postulate 11) states: If two lines are cut by a transversal, the corresponding angles are congruent. Note that single or double arrowheads on the lines identify them as being parallel.

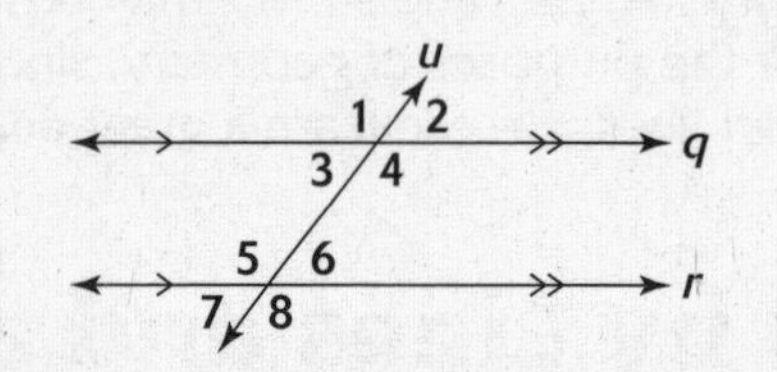

That means that in the preceding figure, $\angle 1$ is congruent to $\angle 5$, $\angle 2$ is congruent to $\angle 6$, $\angle 3$ is congruent to $\angle 7$, and $\angle 4$ is congruent to $\angle 8$. Various theorems can be proven based upon the parallel postulate:

> **Theorem 12:** If parallel lines are cut by a transversal, their alternate interior angles are congruent.
>
> **Theorem 13:** If parallel lines are cut by a transversal, their alternate exterior angles are congruent.
>
> **Theorem 14:** If parallel lines are cut by a transversal, their consecutive interior angles are supplementary.
>
> **Theorem 15:** If parallel lines are cut by a transversal, their consecutive exterior angles are supplementary.

The symbol || means "is parallel to." In the preceding figure, $q \parallel r$.

Example Problems

These problems show the answers and solutions. Refer to the preceding figure to solve these problems.

1. Suppose $m\angle 2 = 60°$. Find $m\angle 6$.

   ***Answer:* 60°** $\angle 6$ is a corresponding angle to $\angle 2$, so by Postulate 11, they are congruent.

2. Suppose that $m\angle 6 = 50°$. Find $m\angle 3$.

   ***Answer:* 50°** Angles 3 and 6 are alternate interior angles, so by Theorem 12, they are congruent.

3. Suppose $m\angle 7 = 70°$. Find $m\angle 1$.

   ***Answer:* 110°** Angles 7 and 1 are consecutive exterior angles, and therefore, by Theorem 15 they are supplementary.

Work Problems

Use these problems for additional practice. Problems 1–5 refer to the following diagram.

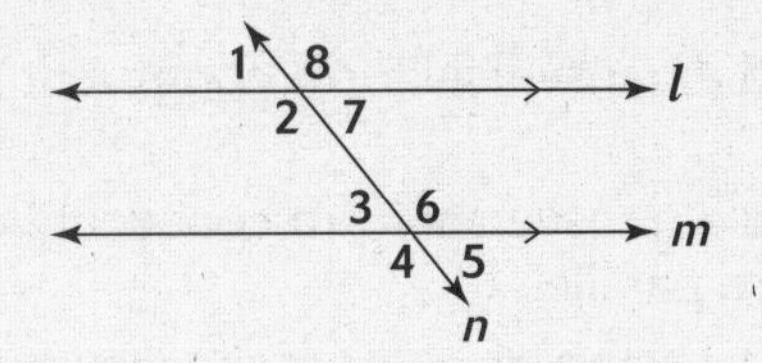

1. If $m\angle 7 = 50°$, what is $m\angle 6$?
2. If $m\angle 2 = 120°$, what is $m\angle 6$?
3. If $m\angle 5 = 45°$, what is $m\angle 8$?
4. If $m\angle 4 = 65°$, what is $m\angle 2$?
5. If $m\angle 3 = 80°$, what is $m\angle 7$?

Worked Solutions

1. **130°** **$\angle 7$ and $\angle 6$ are consecutive interior angles.** By Theorem 14: If parallel lines are cut by a transversal, their consecutive interior angles are supplementary. $180° - 50° = 130°$.
2. **120°** **$\angle 2$ and $\angle 6$ are alternate interior angles.** By Theorem 12: If parallel lines are cut by a transversal, their alternate interior angles are congruent.
3. **135°** **$\angle 5$ and $\angle 8$ are consecutive exterior angles.** By Theorem 15: If parallel lines are cut by a transversal, their consecutive exterior angles are supplementary.
4. **65°** **$\angle 4$ and $\angle 2$ are corresponding angles.** The Parallel Postulate (Postulate 11) states: If two lines are cut by a transversal, the corresponding angles are congruent.
5. **80°** **$\angle 3$ and $\angle 7$ are alternate interior angles.** By Theorem 12: If parallel lines are cut by a transversal, their alternate interior angles are congruent.

Proving Lines Parallel

Postulate 11 and Theorems 12 through 15 tell us that when parallel lines are cut by a transversal, certain angle relationships exist. It is, however, often desirable or necessary to prove two lines to be parallel. With that in mind, it would be very convenient if we could show that the **converses** of those rules were also true. A converse is a postulate or theorem with the *if* and *then* parts reversed. Be sure to note that not all postulates and theorems are reversible. In this case, however, we got lucky. The following diagram illustrates Postulate 12.

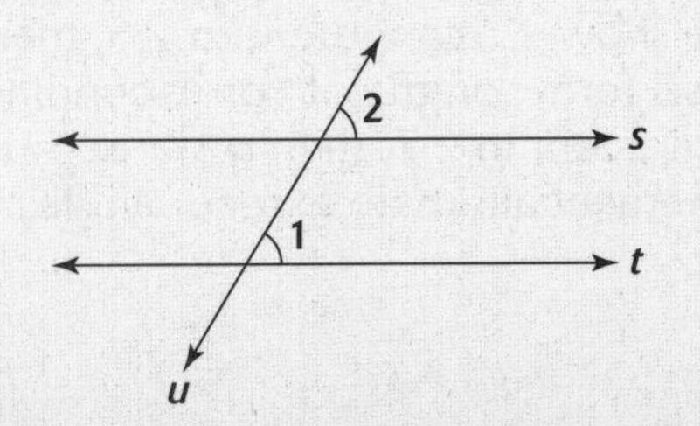

Postulate 12: If two lines and a transversal form congruent corresponding angles, then the lines are parallel.

From that, we are able to prove the converses of the previous theorems.

Theorem 16: If two lines and a transversal form congruent alternate interior angles, then the lines are parallel.

Theorem 17: If two lines and a transversal form congruent alternate exterior angles, then the lines are parallel.

Theorem 18: If two lines and a transversal form consecutive interior angles that are supplementary, then the lines are parallel.

Theorem 19: If two lines and a transversal form consecutive exterior angles that are supplementary, then the lines are parallel.

To these, we can add the ever-popular Theorem 20:

Theorem 20: In a plane, if two lines are perpendicular to the same line, then the lines are parallel.

Its inverse is also true. That is, if a transversal is perpendicular to one of two parallel lines, it must also be perpendicular to the other.

Example Problems

These problems show the answers and solutions. All three questions refer to the following diagram.

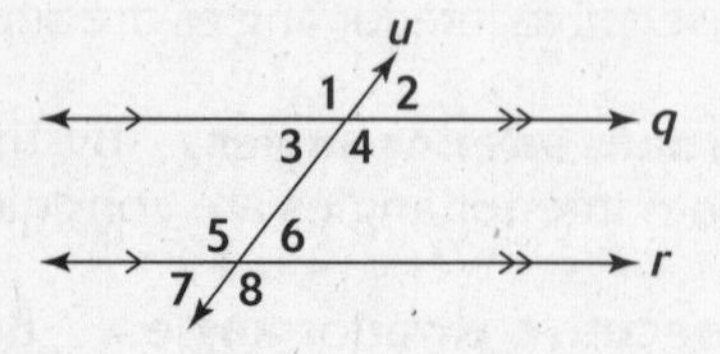

1. To prove $q \parallel r$ what angle must be congruent to $\angle 5$?

   Answer: **$\angle 1$ or $\angle 4$** If we show $\angle 5$ congruent to $\angle 1$, then $q \parallel r$ because of Postulate 12: If two lines and a transversal form congruent corresponding angles, then the lines are parallel. If we show $\angle 5$ congruent to $\angle 4$, then $q \parallel r$ because of Theorem 16: If two lines and a transversal form congruent alternate interior angles, then the lines are parallel.

2. To prove $q \parallel r$ what angle must be supplementary to $\angle 4$?

   Answer: **$\angle 6$** $\angle 4$ can be seen to be supplementary to adjacent $\angle 3$ and $\angle 2$, but that doesn't help. What is needed is for it to be supplementary to $\angle 6$, thereby proving $q \parallel r$ by Theorem 18: If two lines and a transversal form consecutive interior angles that are supplementary, then the lines are parallel.

3. What angle must be congruent to $\angle 2$ in order to prove $q \parallel r$?

   Answer: **$\angle 6$ or $\angle 7$** If we show $\angle 2$ congruent to $\angle 6$, then $q \parallel r$ because of Postulate 12: If two lines and a transversal form congruent corresponding angles, then the lines are parallel. If we show $\angle 2$ congruent to $\angle 7$, then $q \parallel r$ because of Theorem 17: If two lines and a transversal form congruent alternate exterior angles, then the lines are parallel.

Work Problems

Use these problems to give yourself additional practice.

Identify which theorem or postulate would cause lines l and m to be parallel, given the following circumstances. (All five questions refer to the following diagram.)

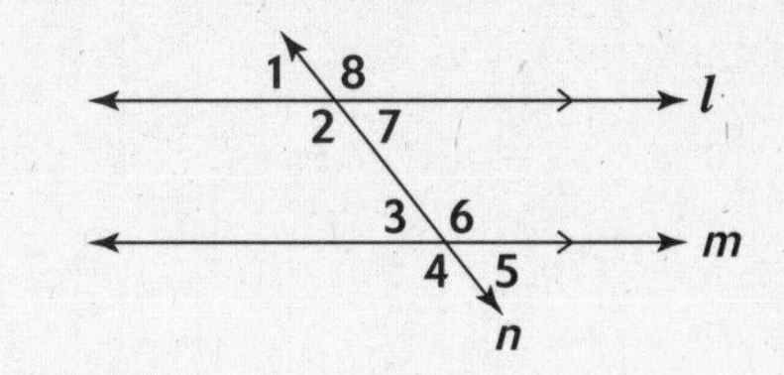

1. $m\angle 8 = m\angle 4$
2. $m\angle 5 = m\angle 7$
3. $m\angle 7 = 55°$; $m\angle 6 = 125°$
4. $m\angle 1 = 50°$; $m\angle 4 = 130°$
5. $m\angle 2 = m\angle 6$

Worked Solutions

1. **Theorem 17** $\angle 8$ and $m\angle 4$ are alternate exterior angles. The appropriate theorem is, therefore, 17: If two lines and a transversal form congruent alternate exterior angles, then the lines are parallel.

2. **Postulate 12** $\angle 5$ and $\angle 7$ are in the same location, below each line and to the right of the transversal. That makes them corresponding angles. If two lines and a transversal form congruent corresponding angles, then the lines are parallel.

3. **Theorem 18** Since $55° + 125° = 180°$, $\angle 7$ and $\angle 6$ are supplementary angles. They are also both on the same side of the transversal and between the lines, so Theorem 18 applies: If two lines and a transversal form consecutive interior angles that are supplementary, then the lines are parallel.

4. **Theorem 19** Since $50° + 130° = 180°$, $\angle 1$ and $\angle 4$ are supplementary angles. They are also both on the same side of the transversal and outside the lines, so Theorem 19 applies: If two lines and a transversal form consecutive exterior angles that are supplementary, then the lines are parallel.

5. **Theorem 16** $\angle 2$ and $\angle 6$ are alternate interior angles, so we use the theorem that pertains to them: If two lines and a transversal form congruent alternate interior angles, then the lines are parallel.

Chapter 3

Triangles

A **triangle** is a closed three-sided plane figure with three interior angles. A triangle is represented by the symbol $\triangle$ and is named by its three **vertices** (plural of vertex). This is $\triangle ABC$, BCA, CAB, etc.

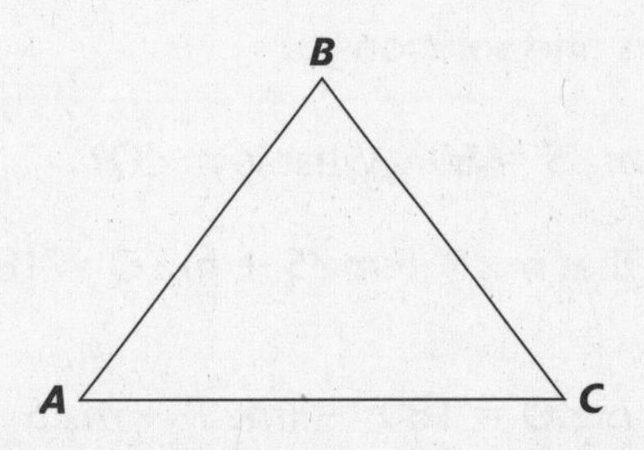

The triangle is the simplest of all closed figures, but one of the most useful. That's because any closed figure with more than three sides can be subdivided into triangles. Here, the five-sided **pentagon** *DEFGH* has been subdivided into $\triangle DEH$, $\triangle EFH$, and $\triangle FGH$.

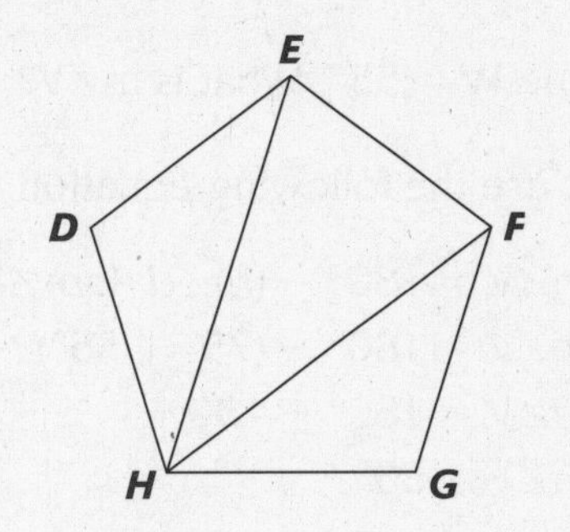

Sum of Angle Measure

The following theorem can be proven with the help of the Parallel Postulate or a pair of scissors:

Theorem 21: The sum of the interior angles of any triangle is 180°.

Following, a triangular shape drawn on paper has been carefully cut, and the three angles have been laid next to one another. Notice that they form a straight angle, which we learned in Chapter 1 has a measure of 180°.

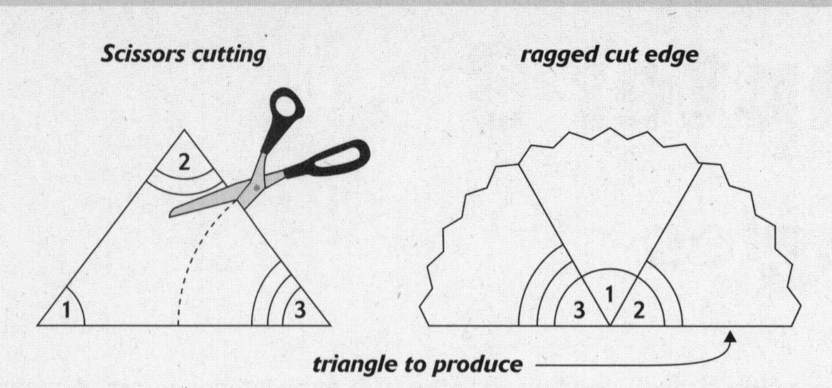

Example Problems

These problems show the answers and solutions.

1. In $\triangle QRS$, $m\angle R = 30°$ and $m\angle S = 50°$. What is $m\angle Q$?

   Answer: 100° We know that $m\angle R + m\angle S + m\angle Q = 180°$. To find $m\angle Q$, use the following equation:

   $$m\angle Q = 180° - (m\angle R + m\angle S)$$
   $$m\angle Q = 180° - (30° + 50°)$$
   $$m\angle Q = 180° - 80°$$
   $$m\angle Q = 100°$$

2. In $\triangle UVW$, $m\angle U = 75°$ and $m\angle W = 55°$. What is $m\angle V$?

   Answer: 50° To find $m\angle V$, use the following equation:

   $$m\angle V = 180° - (m\angle U + m\angle W)$$
   $$m\angle V = 180° - (75° + 55°)$$
   $$m\angle V = 180° - 130°$$
   $$m\angle V = 50°$$

3. In $\triangle LMN$, $m\angle L = 30°$ and $m\angle M = 60°$. What is $m\angle N$?

   Answer: 90° To find $m\angle N$, use the following equation:

   $$m\angle N = 180° - (m\angle L + m\angle M)$$
   $$m\angle N = 180° - (30° + 60°)$$
   $$m\angle N = 180° - 90°$$
   $$m\angle N = 90°$$

Exterior Angles

When one side of a triangle is extended so as to form an angle supplementary to an interior angle of the triangle, the new angle formed is known as an **exterior angle**. $\angle ABD$ is an exterior angle to $\angle ABC$ in $\triangle ABC$.

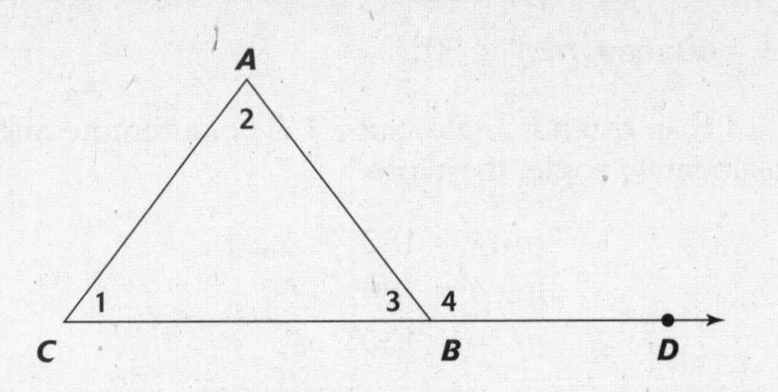

Since $m\angle 1 + m\angle 2 + m\angle 3 = 180°$, and $m\angle 4 + m\angle 3 = 180°$, it is readily provable that $m\angle 4 = m\angle 1 + m\angle 2$. This fact is stated in the form of a theorem.

Theorem 22: An exterior angle of a triangle is equal to the sum of the remote interior angles. ("Remote" means far away, as are angles 1 and 2 in the preceding diagram, as distinguished from the adjacent angle, $\angle 3$.)

Example Problems

These problems show the answers and solutions. Refer to the following figure to solve these problems.

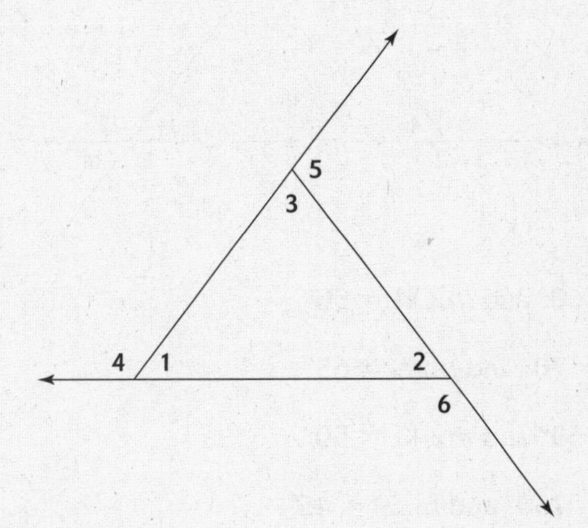

1. Find $m\angle 6$ if $m\angle 1 = 50°$ and $m\angle 3 = 70°$.

   ***Answer:* 120°** Because $\angle 6$ is an exterior angle:

   $$m\angle 6 = m\angle 1 + m\angle 3$$
   $$m\angle 6 = 50° + 70°$$
   $$m\angle 6 = 120°$$

2. Find $m\angle 5$ if $m\angle 1 = 45°$ and $m\angle 2 = 35°$.

   ***Answer:* 80°** Because $\angle 5$ is an exterior angle:

   $$m\angle 5 = m\angle 1 + m\angle 2$$
   $$m\angle 5 = 45° + 35°$$
   $m\angle 5 = 80°$ (even though it is not drawn that way)

3. Find $m\angle 4$ if $m\angle 3 = 60°$ and $m\angle 1 = 50°$.

Answer: 130° $\angle 4$ is an exterior angle, but $\angle 1$ is *not* a remote interior angle. In fact, it's an adjacent supplementary angle, therefore:

$$m\angle 4 = 180° - m\angle 1$$
$$m\angle 4 = 180° - 50°$$
$$m\angle 4 = 130°$$

Work Problems

Use these problems to give yourself additional practice. Refer to the following figure to solve these problems.

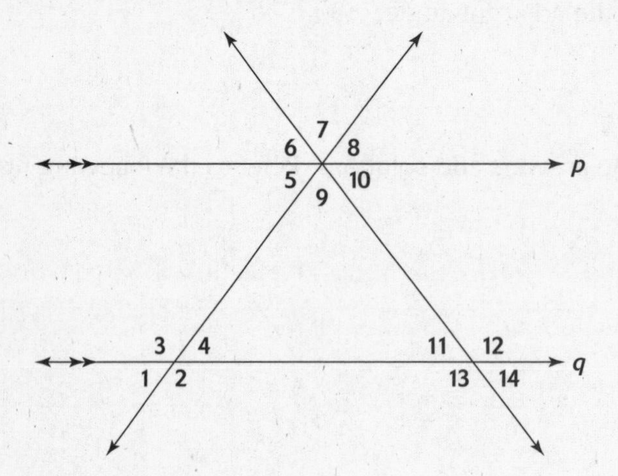

1. Find $m\angle 4$ if $m\angle 9 = 60°$ and $m\angle 11 = 50°$.
2. Find $m\angle 13$ if $m\angle 4 = 70°$ and $m\angle 9 = 65°$.
3. Find $m\angle 3$ if $m\angle 9 = 55°$ and $m\angle 11 = 50°$.
4. Find $m\angle 11$ if $m\angle 2 = 130°$ and $m\angle 9 = 45°$.
5. Find $m\angle 3$ if $m\angle 6 = 60°$ and $m\angle 9 = 65°$.

Worked Solutions

1. **70°** The sum of the interior angles of a triangle is 180°.

$$m\angle 4 = 180° - (m\angle 9 + m\angle 11)$$
$$m\angle 4 = 180° - (60° + 50°)$$
$$m\angle 4 = 180° - 110°$$
$$m\angle 4 = 70°$$

2. **135°** Since $\angle 13$ is an exterior angle:

$$m\angle 13 = m\angle 4 + m\angle 9$$
$$m\angle 13 = 70° + 65°$$
$$m\angle 13 = 135°$$

3. **105°** Since $\angle 3$ is an exterior angle:

$$m\angle 3 = m\angle 9 + m\angle 11$$
$$m\angle 3 = 55° + 50°$$
$$m\angle 3 = 105°$$

4. **85°** Since $\angle 2$ is an exterior angle:

$$m\angle 2 = m\angle 9 + m\angle 11$$
$$130° = 45° + m\angle 11$$
$$m\angle 11 = 130° - 45°$$
$$m\angle 11 = 85°$$

5. **125°** Since $\angle 3$ is an exterior angle:

$$m\angle 3 = m\angle 9 + m\angle 11$$

Since p and q are parallel lines:

$$m\angle 11 = m\angle 6 = 60°$$
$$m\angle 3 = 65° + 60°$$
$$m\angle 3 = 125°$$

Classifying Triangles by Angles

Triangles may be classified according to their sides or their angles. First, let's try classifying by angles. A triangle with all interior angles acute is called an **acute triangle**. Here's an example of an acute triangle:

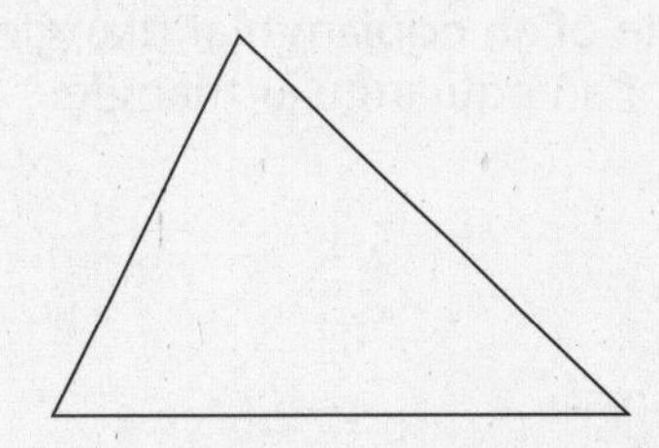

A triangle that contains one interior right angle is called a **right triangle**. Following is an example of a right triangle:

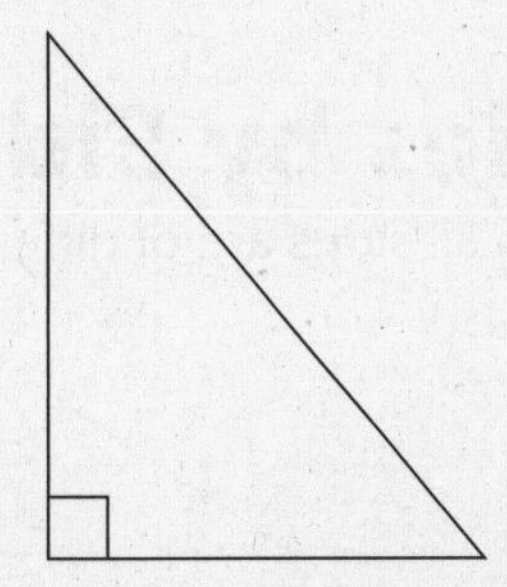

A triangle that contains one interior obtuse angle is called an **obtuse triangle**. This diagram shows an example of an obtuse triangle:

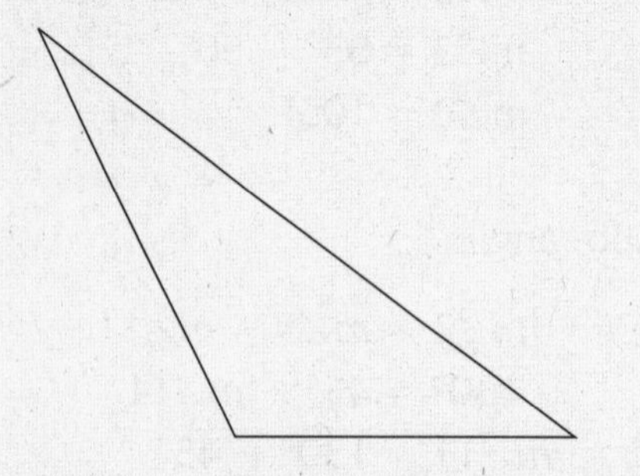

A triangle in which two interior angles are of equal measure is called an **isosceles triangle**. An isosceles triangle may be acute, right, or obtuse. Here is an example of all three types of isosceles triangle:

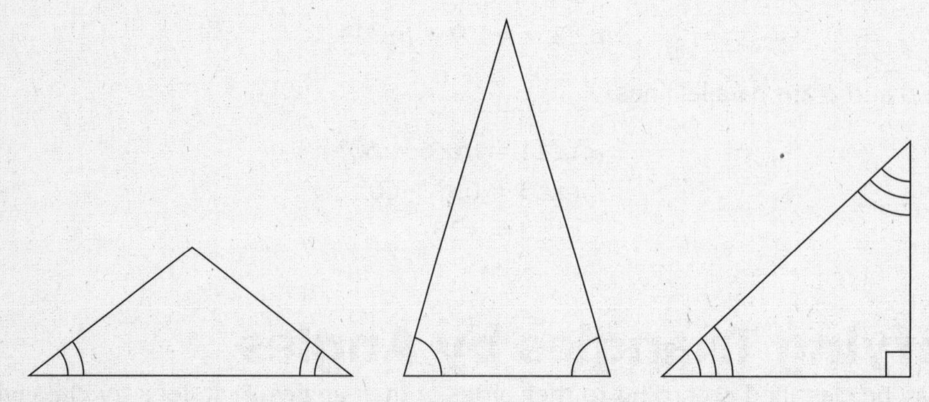

A triangle in which all three interior angles are of equal measure is called an **equiangular triangle**. An equiangular triangle is always acute and adheres to Theorem 23.

> **Theorem 23:** Each interior angle of an equiangular triangle has a measure of 60°. Do you see why? Here is an example of an equiangular triangle:

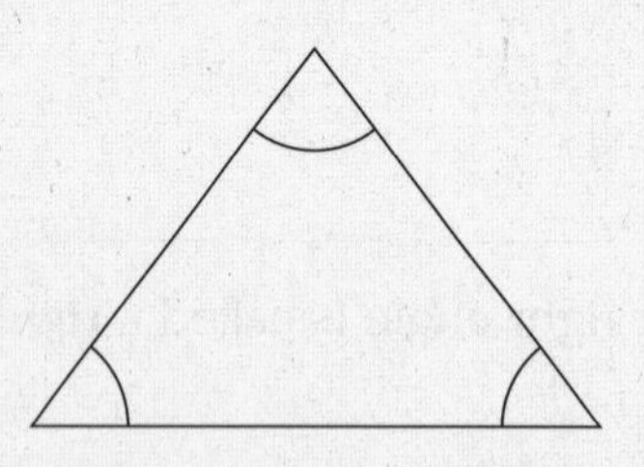

Classifying Triangles by Sides

A **scalene triangle** is a triangle in which all sides are of different length. This is a scalene triangle:

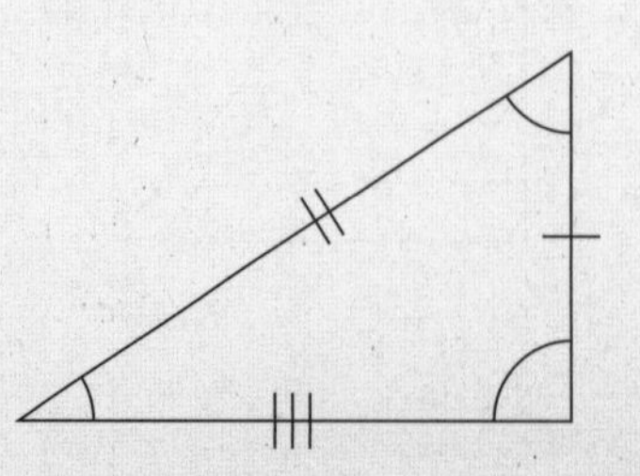

A triangle in which two sides are equal in length is known as an **isosceles triangle**. Does that sound familiar? That's right, an isosceles triangle has two equal sides, as well as two equal angles that are opposite, or across from, those equal sides. The following triangle has both equal pairs marked as such:

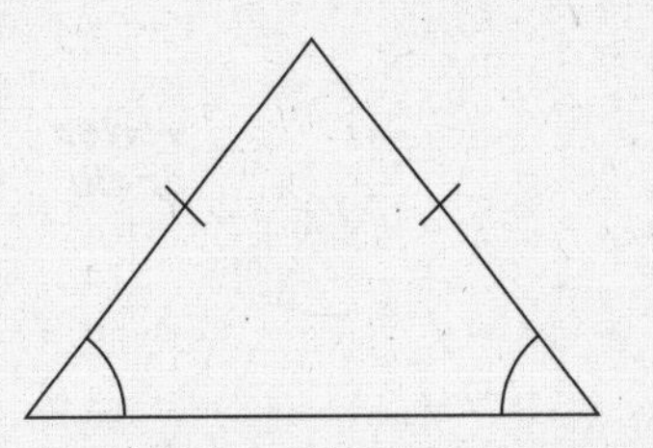

A triangle in which all sides are equal, or congruent, is known as an **equilateral triangle**. An equilateral triangle is also equiangular, and vice versa.

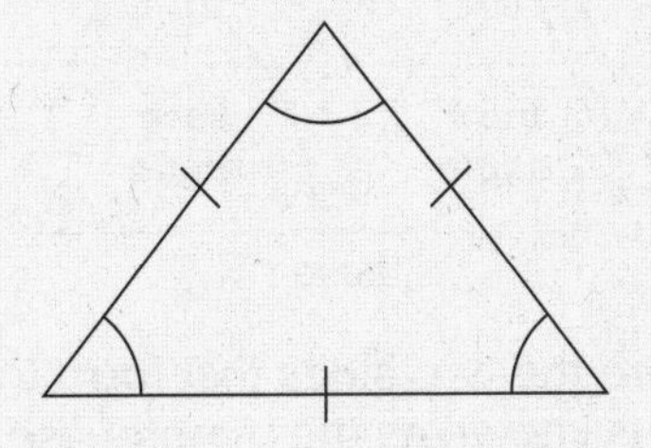

Example Problems

These problems show the answers and solutions. Use the following figure to solve all problems.

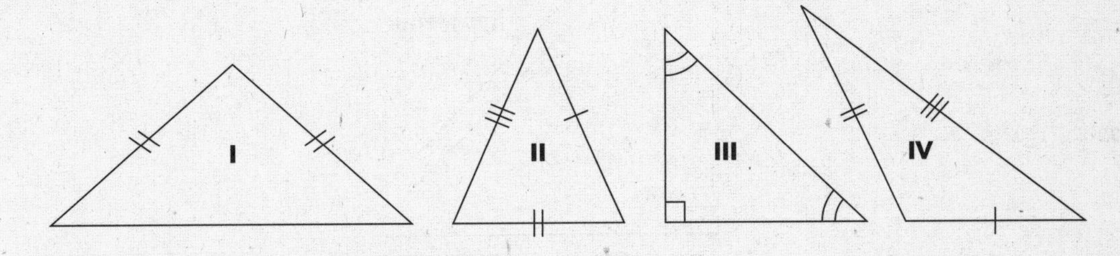

1. Which triangle(s) is/are acute?

   ***Answer:* II** In an acute triangle, all angles must be less than 90°. Only △II passes this test. If you need to prove that to yourself, try fitting the corner of a piece of paper into each angle on △II, and you'll see that it obscures one side of the angle, thus proving each to be less than 90°. Perform the same test in the other triangles if you like, and you'll see that in one angle of each that test doesn't work.

2. Which triangle(s) is/are obtuse?

   ***Answer:* I, IV** An obtuse triangle must contain one angle greater than 90°. Try to fit the corner of a sheet of paper into the topmost (apex) angle of △I, and you'll see its measure is larger than 90°. The angle on the left bottom of △IV will also pass this test.

3. Which triangle(s) is/are isosceles?

   ***Answer:* I, III** An isosceles triangle has two equal angles and two equal sides, but it is necessary to show only one of those pairings to prove that the triangle is isosceles. In △I, two sides are marked as being equal. In △III, two angles are so marked.

Specially Named Sides and Angles

In an isosceles triangle, the two equal sides are called the **legs**, and the third side is called the **base**. The angles formed by each leg and the base are called **base angles**. The remaining angle is known as the **vertex angle**.

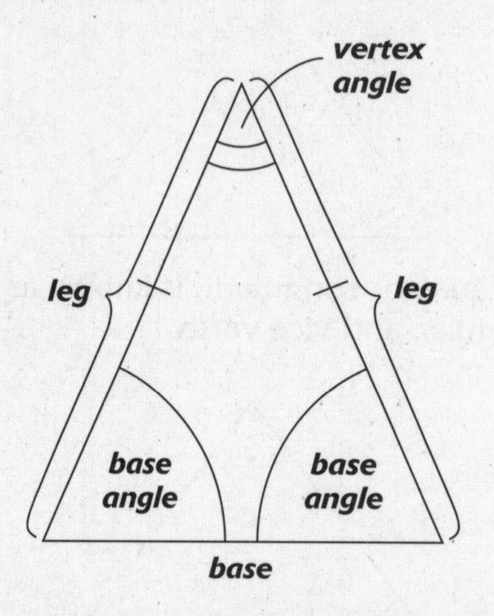

A right triangle also has legs. They are the two sides that form the 90° angle. The third side (opposite, or across from, the right angle) is known as the hypotenuse (pronounced hi POT 'n youss).

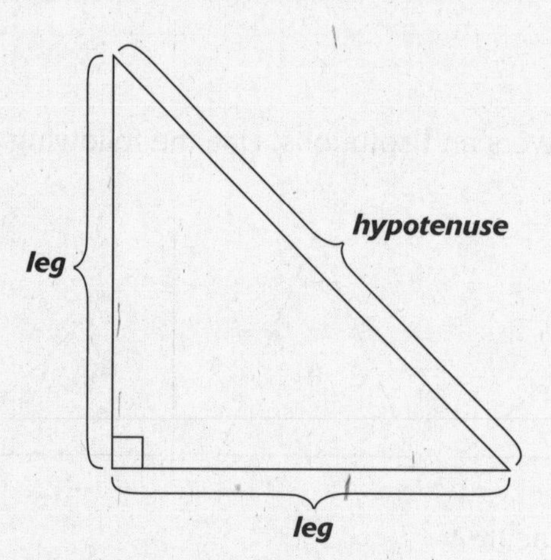

Work Problems

Use these problems to give yourself additional practice. All problems refer to the following figure.

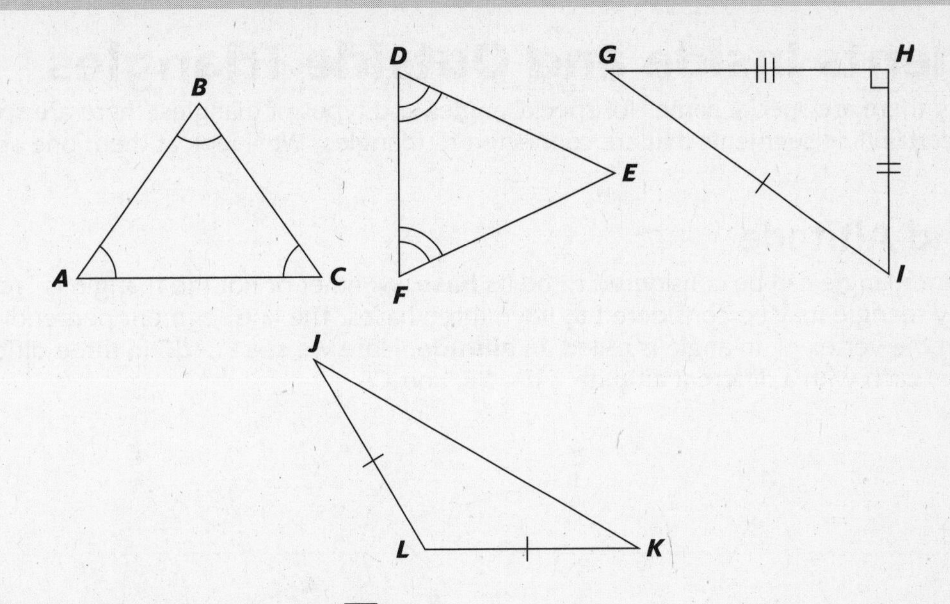

1. $\overline{AB}$ is 5 cm long. How long is $\overline{BC}$?

2. Which triangle of those in the preceding figure is scalene?

3. $\overline{DE}$ is 7 inches long. What is the length of $\overline{DF}$?

4. $m\angle J = 25°$. Find $m\angle L$.

5. Name the only hypotenuse in the figure.

Worked Solutions

1. **5 cm** $\triangle ABC$ is marked as equiangular, which makes it also equilateral. All sides are the same length.

2. **$\triangle GHI$** In a scalene triangle, all sides are of different lengths. Only $\triangle GHI$ fits that requirement as shown by the different markings on its sides.

3. **Unknown** We know that $\overline{DE}$ and $\overline{EF}$ are each 7 inches long, since they are the legs of isosceles $\triangle DEF$. We also know that $\overline{DF}$ is greater than 0 inches and less than 14 inches, or else $\triangle DEF$ could not exist.

4. **130°** $\angle J$ and $\angle K$ are the base angles of isosceles $\triangle JKL$. That means $m\angle J = m\angle K$. Therefore:

$$m\angle L = 180° - (m\angle K + m\angle J)$$
$$m\angle L = 180° - (25° + 25°)$$
$$m\angle L = 180° - 50°$$
$$m\angle L = 130°$$

5. **$\overline{GI}$** $\triangle GHI$ is the only right triangle in the group. The hypotenuse is the side opposite the right angle. That's $\overline{GI}$.

Segments Inside and Outside Triangles

Just the way there are special names for special angles and types of triangles, there are special names for certain line segments that are connected to triangles. We'll look at them one at a time.

Base and Altitude

Any side of a triangle can be considered to be its **base**, whether or not the triangle is "resting on it," so every triangle may be considered to have three bases. The line segment perpendicular to a base from the vertex of an angle is called an **altitude**. Here we see $\triangle ABC$ in three different orientations, each with a different altitude, $\overline{AD}$, $\overline{BE}$, and $\overline{CF}$.

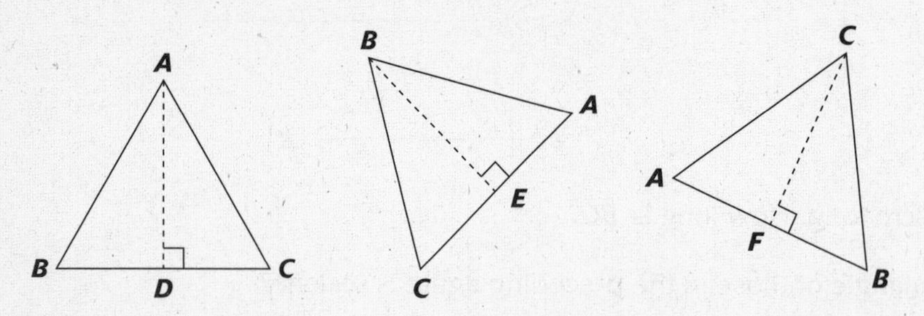

In fact, every triangle has three altitudes, and they do not necessarily fall inside the triangle. Look at this right triangle:

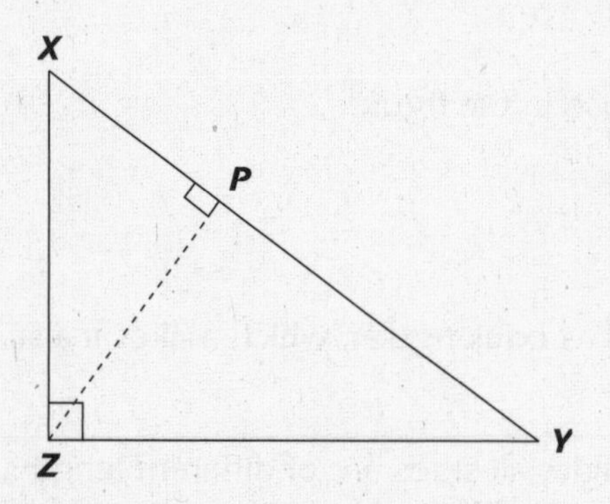

Each leg could serve as an altitude, so $\overline{XZ}$ is the altitude to base $\overline{ZY}$, and $\overline{ZY}$ is the altitude to base $\overline{XZ}$. Of course, there is also altitude $\overline{ZP}$ to base $\overline{XY}$.

In the case of an obtuse triangle, two of the altitudes fall completely outside of the triangle.

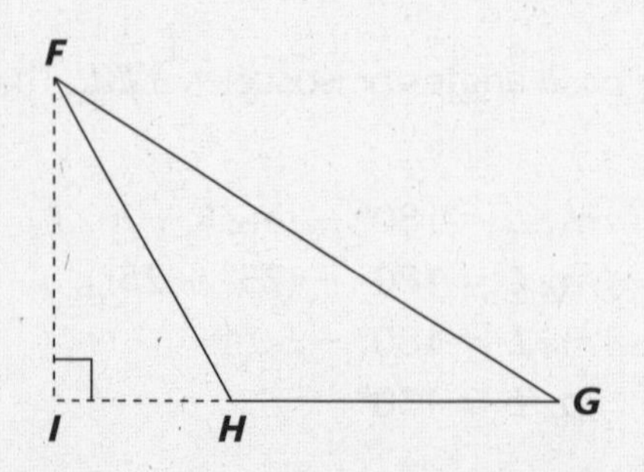

$\overline{HI}$ is an extension of base $\overline{HG}$. It is noteworthy that all three altitudes (or the lines that contain them) intersect at a single point, which may be inside or outside of the triangle.

Median

A triangle's **medians** are the line segments that may be drawn from any vertex to the midpoint of the opposite side.

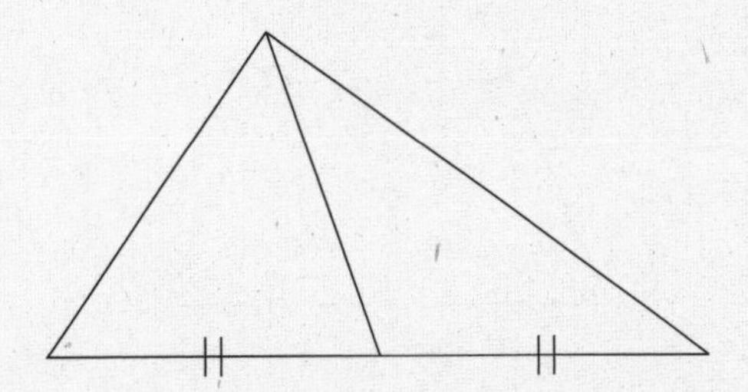

Every triangle has three medians. In every triangle, all three medians intersect at a single point inside the triangle.

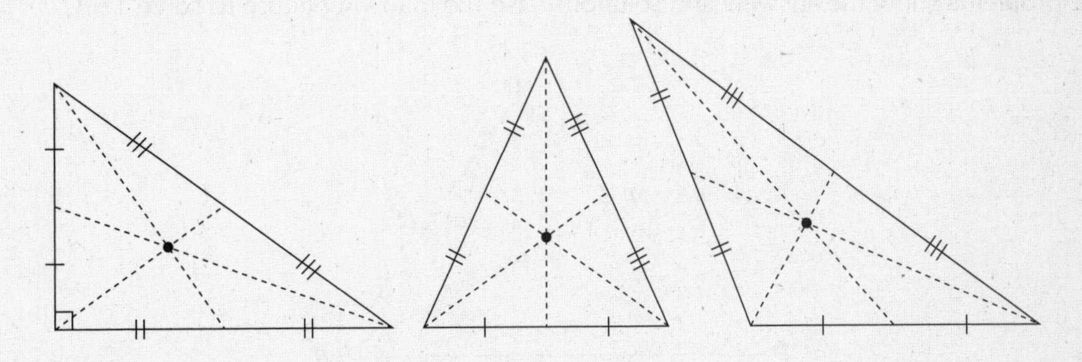

Angle Bisector

An **angle bisector** is a segment drawn from any vertex of a triangle that cuts that angle into two equal angles.

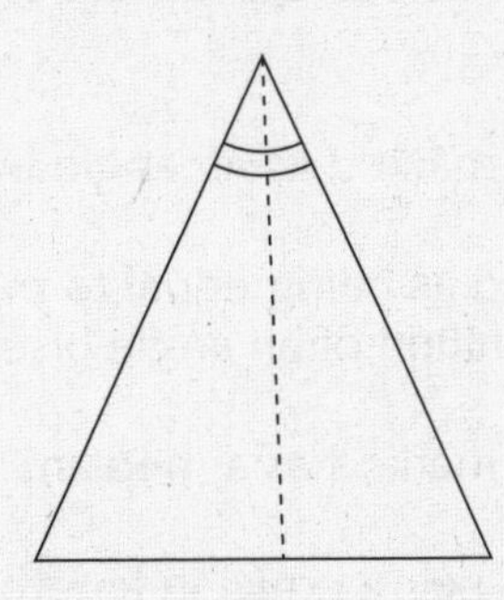

In every triangle, the three angle bisectors intersect at a single point inside the triangle.

As a general rule, the median, altitude, and angle bisector are three different line segments. There are exceptions, however. The altitude from the vertex angle to the base of an isosceles triangle can be proven to be the angle bisector of that angle, as well as the median to the base.

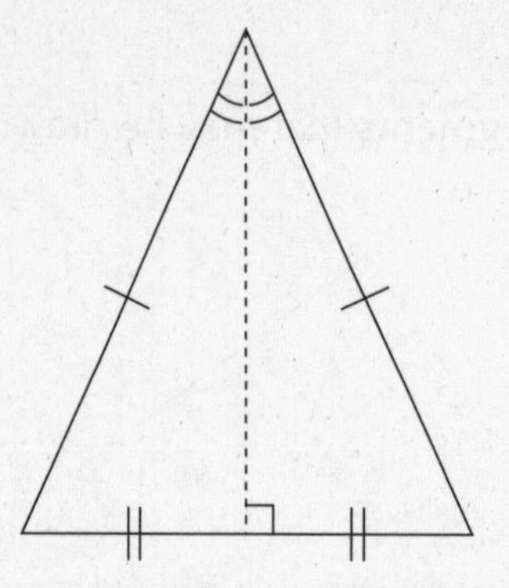

Example Problems

These problems show the answers and solutions. Use the following figure to solve 1–3.

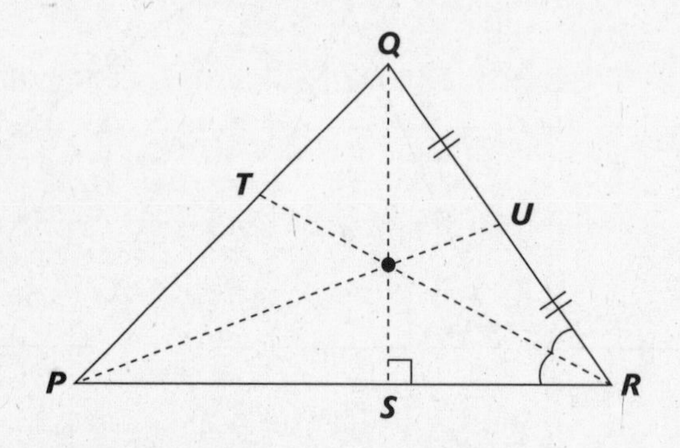

1. In $\triangle PQR$, which line segment is marked as an altitude?

   Answer: $\overline{QS}$ The $\perp$ sign at *S* indicates $\overline{QS}$ forms a right angle with $\overline{PR}$. That's the definition of an altitude.

2. In $\triangle PQR$, which line segment is marked as an angle bisector?

   Answer: $\overline{RT}$ $m\angle QRT$ is marked as being equal to $m\angle PRT$. The segment that cuts $\angle Q$ into two equal angles is the definition of an angle bisector.

3. In $\triangle PQR$, which line segment is marked as a median?

   Answer: $\overline{PU}$ $\overline{QU}$ is marked as being equal in length to $\overline{RU}$. The line from the vertex that divides the opposite side into two equal segments is the definition of a median.

Congruent Triangles

Congruent means identical in shape and size. **Congruent triangles** are triangles that are identical in shape and size. If you put one congruent triangle over another so that their congruent angles and sides coincided, you would see only one triangle. The symbol $\cong$ means "is congruent to." $\triangle ABC \cong \triangle DEF$ means $\triangle ABC$ is congruent to $\triangle DEF$.

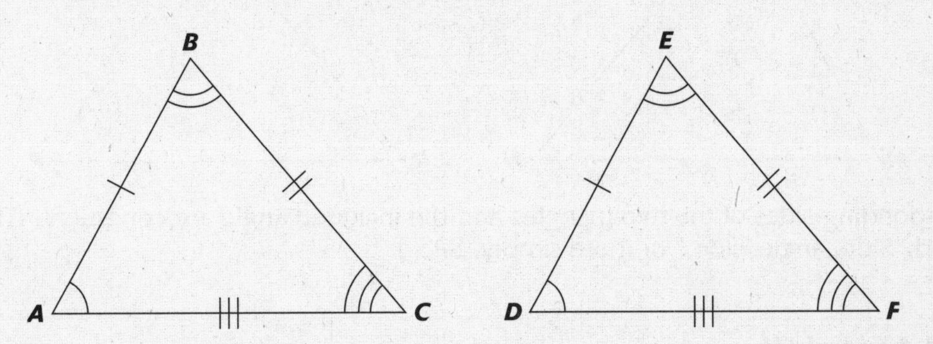

Because $\triangle ABC \cong \triangle DEF$, $AB = DE$, $BC = EF$, $AC = DF$, $m\angle A = m\angle D$, $m\angle B = m\angle E$, $m\angle C = m\angle F$.

Proofs of Congruence

Although in two congruent triangles each side of one must be equal in size to each **corresponding** side of the other (corresponding means the one it matches up with), and each angle of one must be equal in measure to its corresponding angle, it is not necessary to prove that in order to prove two triangles congruent. Several postulates follow that name minimum conditions for two triangles to be congruent:

SSS and SAS

Postulate 13 (SSS): If each side of one triangle is congruent to the corresponding side of a second triangle, then the two triangles are congruent.

$\triangle GHI \cong \triangle JKL$ **by SSS**

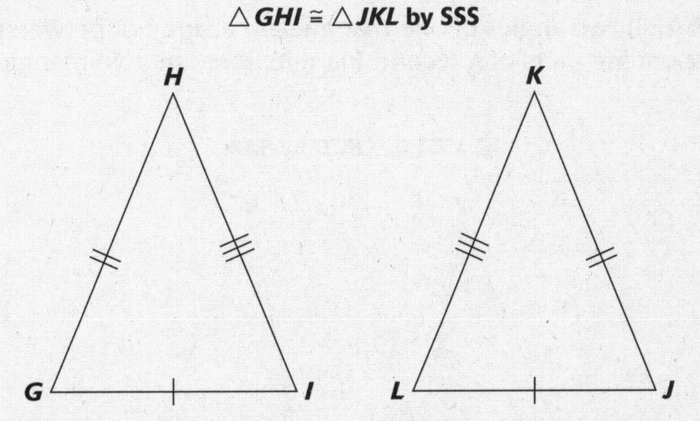

The corresponding sides of the two triangles are congruent. (This is abbreviated "Side, side, side," or more simply, SSS.)

Postulate 14 (SAS): If two sides of one triangle and the angle between them are congruent to the corresponding parts of a second triangle, then the two triangles are congruent.

$\triangle MNO \cong \triangle PRQ$ by SAS

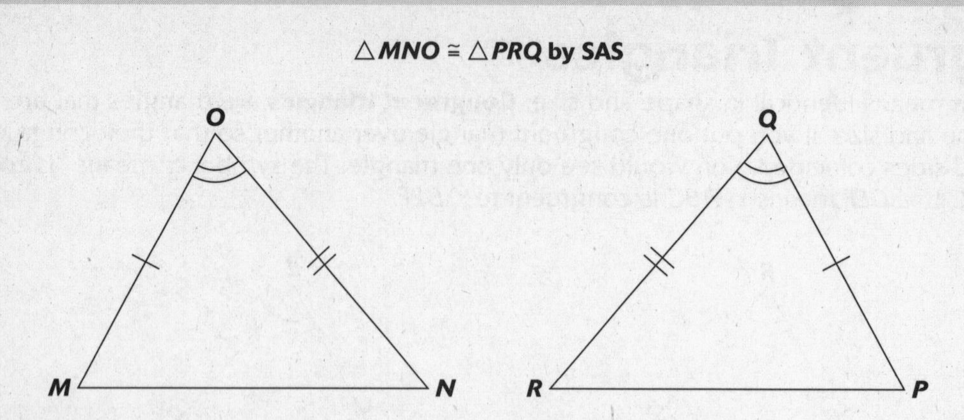

Two corresponding sides of the two triangles and the included angle are congruent. (This is abbreviated "Side, angle, side," or more simply, SAS.)

ASA, SAA, and HL

Postulate 15 (ASA): If two angles of one triangle and the side between them are congruent to the corresponding parts of a second triangle, then the two triangles are congruent.

$\triangle STU \cong \triangle VWX$ by ASA

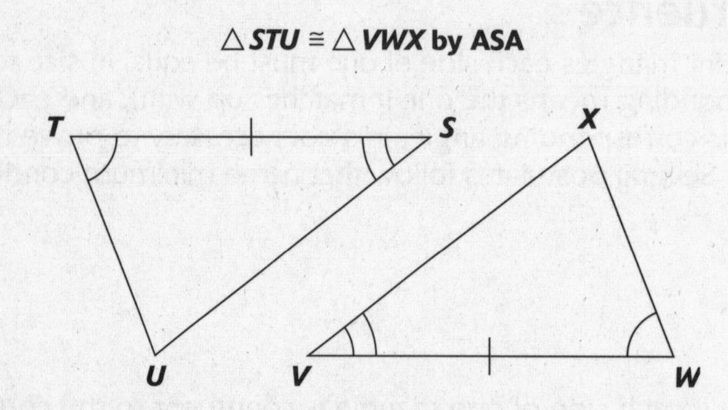

Two corresponding angles of the two triangles and the included side are congruent. (This is abbreviated "Angle, side, angle," or more simply, ASA.)

Postulate 16 (SAA): If two angles of one triangle and a side not between them are congruent to the corresponding parts of a second triangle, then the two triangles are congruent.

$\triangle YZA \cong \triangle BCD$ by SAA

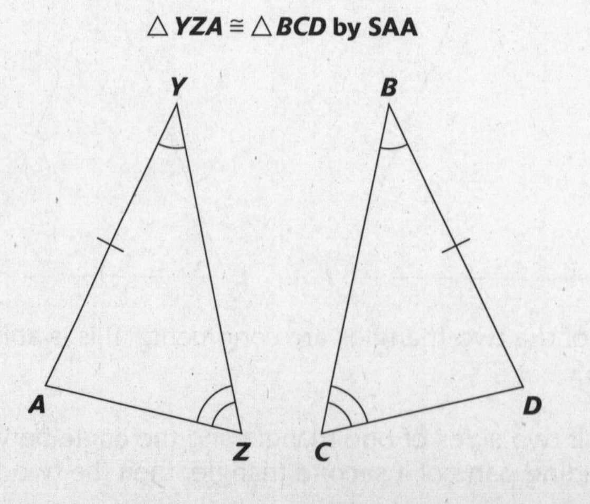

Two corresponding angles of the two triangles and a nonincluded side are congruent. (This is abbreviated "Side, angle, angle," or more simply, SAA.) Some books write this as "angle, angle, side," or AAS. Whatever you do, don't confuse it with "angle, side, side."

Postulate 17 (HL): If the hypotenuse and one leg of one right triangle are congruent to the corresponding parts of a second right triangle, then the two right triangles are congruent.

$\triangle EFG \cong \triangle HIJ$ **by HL**

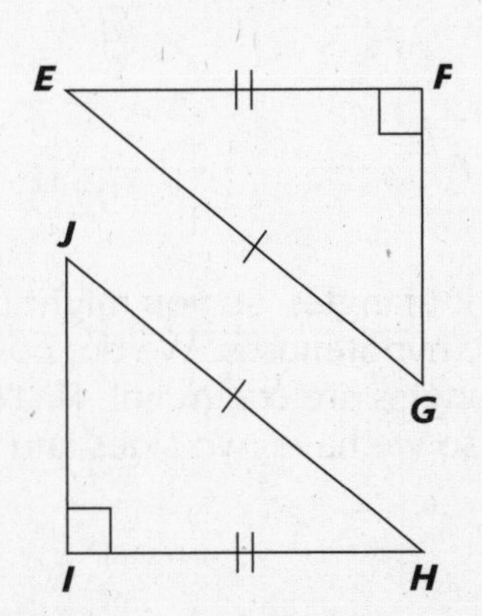

The hypotenuse and one leg of one right triangle are congruent to the corresponding parts of a second right triangle. (This is abbreviated "Hypotenuse leg," or more simply, HL.)

Example Problems

These problems show the answers and solutions. For all four questions, tell the reason why the triangles in each diagram are congruent (if they are).

1.

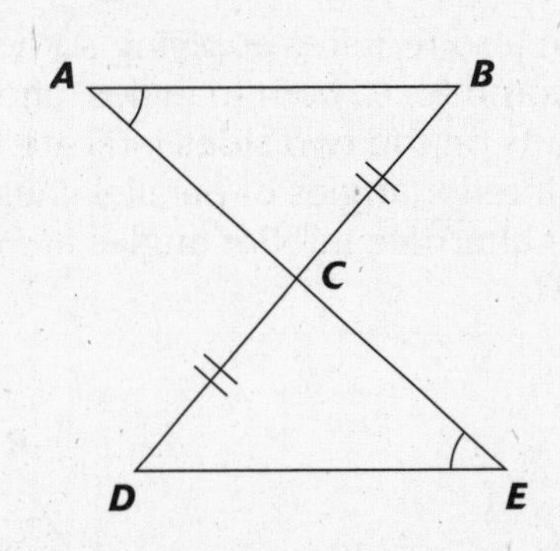

***Answer:* SAA** $\overline{BC}$ is marked congruent to $\overline{CD}$, so that's a side. $\angle A$ is marked as congruent to $\angle E$ (a shortcut for saying their angle measures are equal); that's a side and an angle. We know nothing about angles *D* and *B* or the remaining two sides. How are $\angle ACB$ and $\angle DCE$ related? Remember back to Chapter 1 where we looked at vertical angles. Theorem 7 tells us that vertical angles are equal in measure. That's all we need to apply SAA.

2.

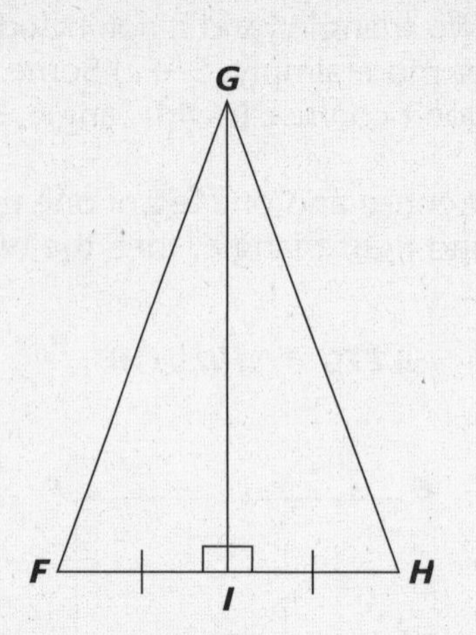

***Answer:* SAS** These are right triangles, so you might think the rule must be HL, but we have no information about the hypotenuses. We do, however, know that $\overline{FI} \cong \overline{HI}$. Also, $\angle GIF \cong \angle GIH$, since all right angles are congruent. Finally, in case you hadn't noticed, $\overline{GI}$ is common to both triangles, so we have two sides and the included angle being congruent. That's SAS.

3.

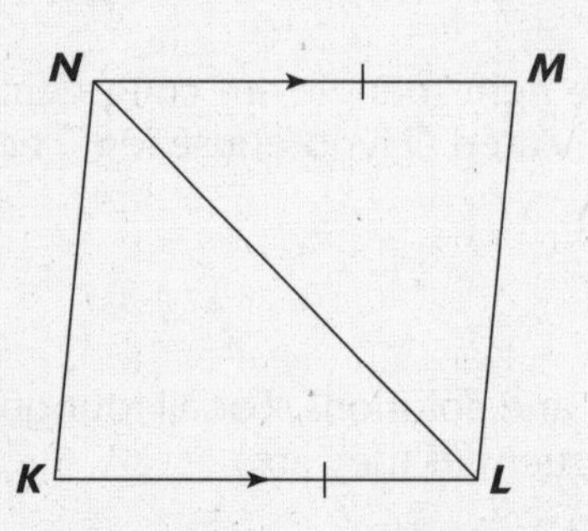

***Answer:* SAS** This problem also requires applying some previously learned facts—geometry is like that. $\overline{NL}$ is common to both triangles, and $\overline{NM}$ is shown congruent to $\overline{KL}$. $\overline{NM} // \overline{KL}$ also (the arrowheads on the two sides indicate their parallel quality). That makes $\angle MNL$ and $\angle KLN$ alternate interior angles of parallel lines. By Theorem 12, if parallel lines are cut by a transversal, their alternate interior angles are congruent. Hence, we have SAS as the reason $\triangle LMN \cong \triangle NKL$.

4.

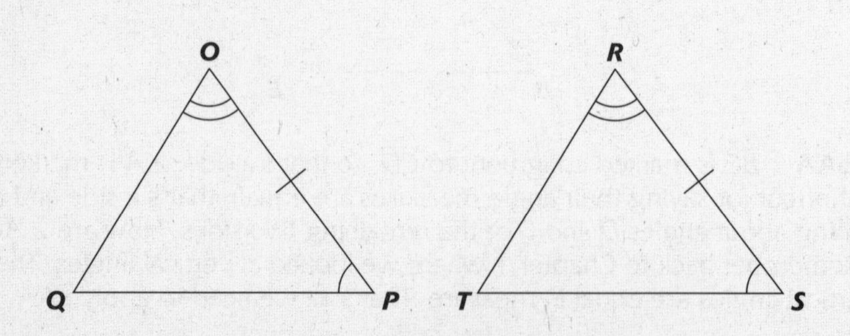

***Answer:* ASA** This one is perfectly straightforward as $m\angle O = m\angle R$, $m\angle P = m\angle S$, and $\overline{OP} \cong \overline{RS}$, so two angles and the included side match up.

Right Triangle Specials

The hypotenuse-leg congruency postulate (17) is unique among all congruency postulates, since two sides alone are mentioned, with their positions relative to the (right) angle not considered. There are other special right triangle proofs of congruency (such as HA, LL, and LA), but each is a special form of one that you should already know.

Theorem 24 (HA): If the hypotenuse and an acute angle of one right triangle are congruent to the corresponding parts of a second right triangle, then the two triangles are congruent.

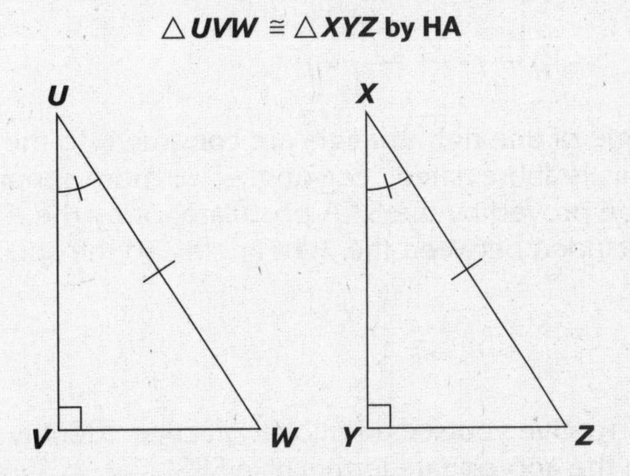

The hypotenuse and one angle of one right triangle are congruent to the corresponding parts of a second right triangle. (This is abbreviated "Hypotenuse angle," or more simply, HA.) Notice that, in reality, this theorem is proved by the SAA postulate, since we know that each triangle has a corresponding right angle.

Theorem 25 (LL): If the legs of one right triangle are congruent to the corresponding parts of a second right triangle, then the two triangles are congruent.

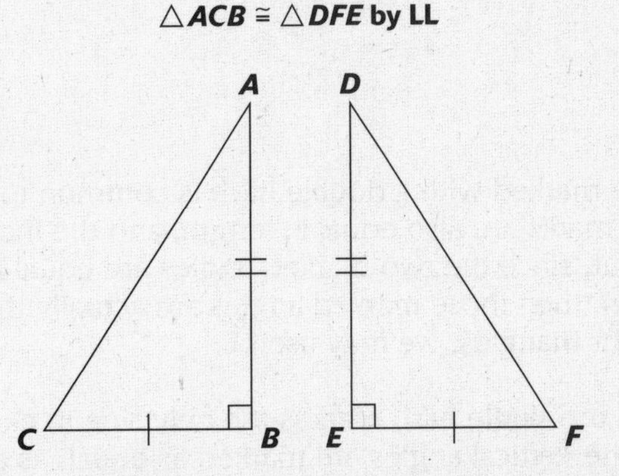

The legs of one right triangle are congruent to the corresponding legs of a second right triangle. (This is abbreviated "leg leg," or more simply, LL.) Notice that, in reality, this theorem is proved by the SAS postulate, with the corresponding right angle being the included angle.

Theorem 26 (LA): If one leg and an acute angle of one right triangle are congruent to the corresponding parts of a second right triangle, then the two triangles are congruent.

$\triangle GHI \cong \triangle JLK$ **by LA**

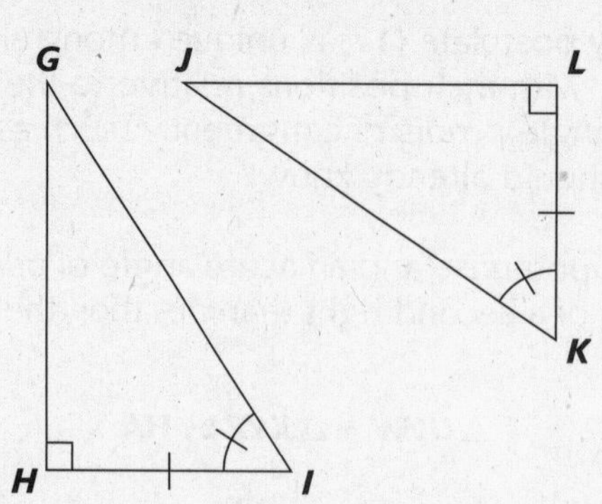

One leg and one acute angle of one right triangle are congruent to the corresponding parts of a second right triangle. (This is abbreviated "Leg angle," or more simply, LA.) Notice that, in reality, this theorem may be proved by the SAA postulate, or by the ASA postulate, depending upon whether the leg is included between the right angle and the specified acute one.

Work Problems

Use problems 1 through 5 to give yourself additional practice. Identify the reason each pair of triangles is congruent. Use the appropriate letters (like SSS).

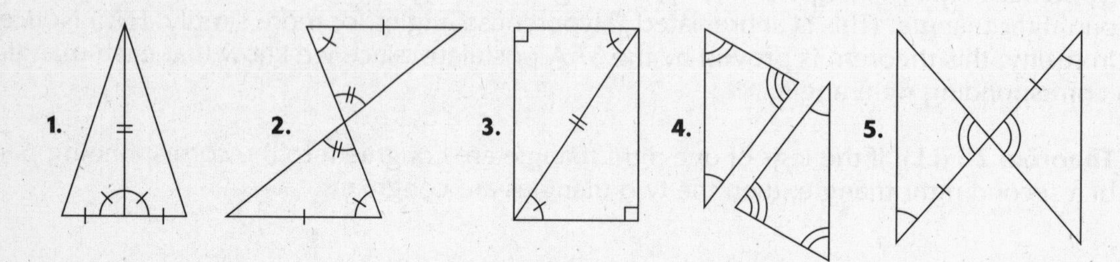

Worked Solutions

1. **SAS or LL** The side marked with a double hash is common to both triangles. The sides with the single hash marks are also equal in length, and the included angles are marked as equal, hence SAS. But, since the two marked angles are equal and supplementary (they form a straight angle), then those marked angles are actually right angles. Since we're dealing with two right triangles, we may use LL.

2. **SAA** The side with the single hash mark is the only side in each triangle we have any information about. The vertical angles are marked as equal, as are the other corresponding pair, so use SAA.

3. **HA or SAA** Both are right triangles, with acute angles marked as equal and with a shared hypotenuse. That allows the HA reason. Since the right angles are also of equal measure (of course), SAA is also a viable choice.

4. **None** AAA is not a proof of congruence. No information is given about the sides, so there is no provable congruency here.

5. **ASA** The side of each triangle marked with a single hash mark is included between the marked angles of equal measure. That's ASA. SAA, however, would not have been wrong; just less desirable when another choice is available.

Corresponding Parts (CPCTC)

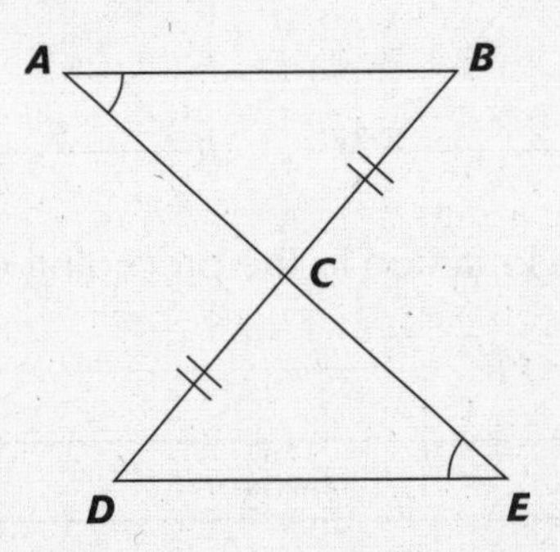

In the preceding figure, $\angle A \cong \angle E$, $\angle B \cong \angle D$, and $\angle ABC \cong \angle EDC$. Additionally, $\overline{BC} \cong \overline{CD}$, $\overline{AB} \cong \overline{DE}$, and $\overline{AC} \cong \overline{CE}$. Assuming that $\triangle ABC \cong \triangle EDC$, each pair of angles and sides just named is congruent, since **corresponding parts of congruent triangles are congruent**. This mouthful may be handily abbreviated as **CPCTC**.

Now, rather than assuming the congruency of the two triangles we'll prove it formally. By so doing, we'll be able to prove conclusively that $AC \cong CE$.

Example Problems

These problems show the answers and solutions.

1. *Given*: $\triangle ABC$ and $\triangle CDE$, as marked in the previous diagram.

   Prove: $\triangle ABC \cong \triangle EDC$, and $\overline{AC} \cong \overline{CE}$.

Statement	*Reason*
1. $m\angle A = m\angle E$	1. Given
2. $\overline{BC} \cong \overline{CD}$	2. Given
3. $m\angle ACB \cong m\angle DCE$	3. Vertical angles are congruent.
4. $\triangle ABC \cong \triangle EDC$	4. SAA
5. $\overline{AC} \cong \overline{CE}$	5. CPCTC

2.

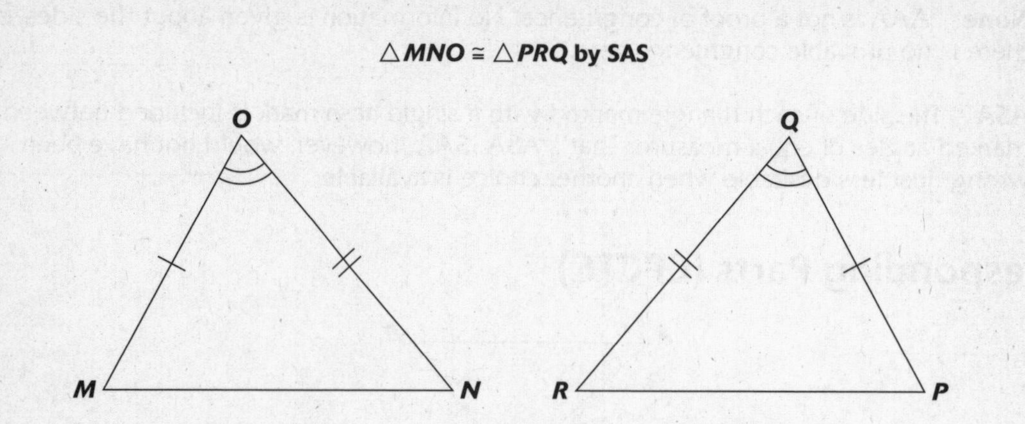

Given: $\triangle MNO$ and $\triangle PQR$, as marked in the preceding diagram.

Prove: $\angle M \cong \angle P$, and $\overline{MN} \cong \overline{PR}$.

Statement	Reason
1. $m\angle O = m\angle Q$	1. Given
2. $\overline{MO} \cong \overline{PQ}$	2. Given
3. $\overline{NO} \cong \overline{QR}$	3. Given
4. $\triangle MNO \cong \triangle PRQ$	4. SAS
5. $\angle M \cong \angle P$, $\overline{MN} \cong \overline{PR}$	5. CPCTC

3.

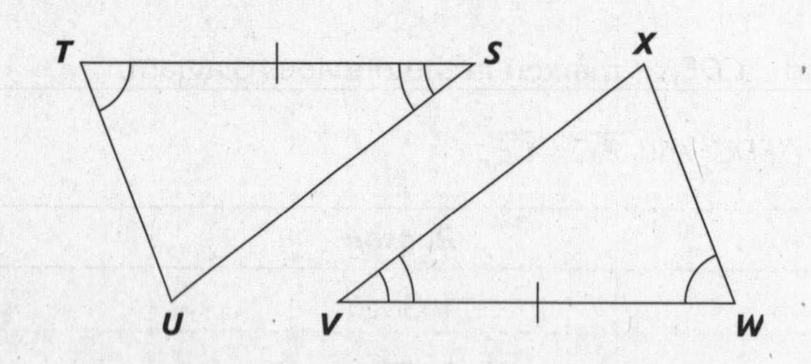

Given: $\triangle STU$ and $\triangle VWX$, as marked in the preceding diagram.

Prove: $\overline{WX} \cong \overline{TU}$, and $\overline{VX} \cong \overline{SU}$.

Statement	Reason
1. $m\angle T = m\angle W$	1. Given
2. $\overline{TS} \cong \overline{VW}$	2. Given
3. $m\angle S = m\angle V$	3. Given
4. $\triangle STU \cong \triangle VWX$	4. ASA
5. $\overline{WX} \cong \overline{TU}$, $\overline{VX} \cong \overline{SU}$	5. CPCTC

Isosceles Triangles

Isosceles triangles are special, and so there are special relationships among their sides and angles. In $\triangle ABC$ in the following figure, median $\overline{AD}$ has been drawn.

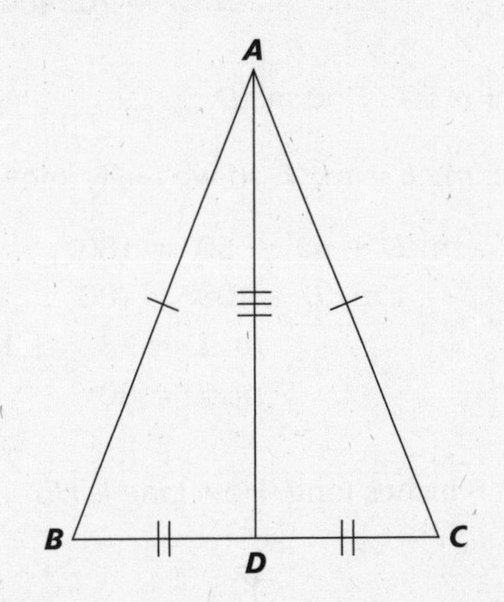

It easily can be proven that $\triangle ABD \cong \triangle ACD$. In turn, that fact leads us to some rather important theorems.

> **Theorem 27:** If two sides of a triangle are congruent, then the angles opposite those sides are congruent.
>
> **Theorem 28:** If two angles of a triangle are congruent, then the sides opposite those angles are congruent.
>
> **Theorem 29:** If a triangle is equilateral, then it is also equiangular.
>
> **Theorem 30:** If a triangle is equiangular, then it is also equilateral.

Example Problems

These problems show the answers and solutions. Use the following diagram to solve problems 1 and 2.

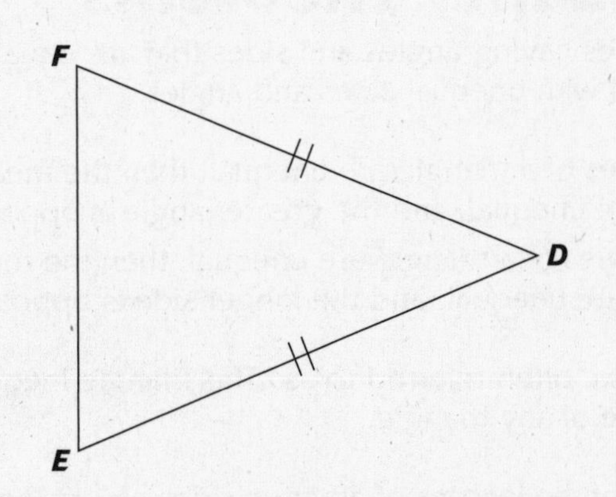

1. $m\angle D$ in the isosceles $\triangle DEF$ is 40°. Find $m\angle F$.

   ***Answer:* 70°** Since $m\angle D + m\angle E + m\angle F = 180°$, and because $ED = DF$, Theorem 27 tells us that $m\angle E = m\angle F$. We can do the following:

$$m\angle D + m\angle E + m\angle F = 180°$$
$$40° + 2(m\angle F) = 180°$$
$$2(m\angle F) = 140°$$
$$m\angle F = 70° \text{ and, by the way, } m\angle E = 70°$$

2. $m\angle E$ in the isosceles $\triangle DEF$ is 50°. Find $m\angle D$.

***Answer:* 80°** Remember, $m\angle E = m\angle F$, so we really know two angles already.

$$m\angle D + 50° + 50° = 180°$$
$$m\angle D + 100° = 180°$$
$$m\angle D = 180° - 100°$$
$$m\angle D = 80°$$

3. In $\triangle GHI$ that follows, GI is 8 inches long. How long is HI?

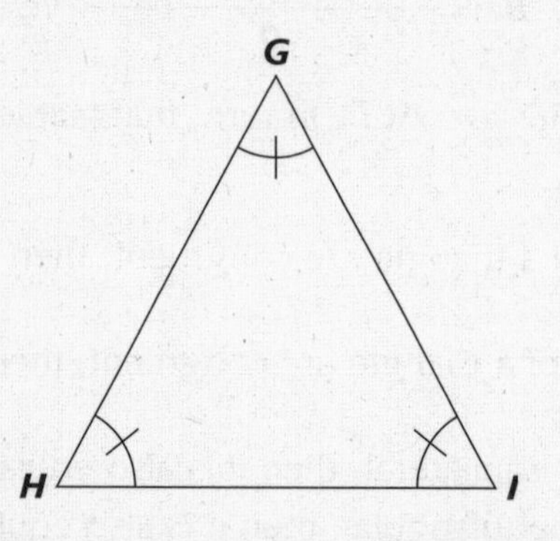

***Answer:* 8 inches** According to Theorem 30, if a triangle is equiangular, then it is also equilateral. In the diagram, $m\angle G = m\angle H = m\angle I$. That means that $HI = GI = HG = 8$ inches.

Triangle Inequality Theorems

We've just dealt with triangles having angles and sides that are equal. Two important theorems are concerned with triangles with unequal sides and angles.

> **Theorem 31:** If two sides of a triangle are unequal, then the measures of the angles opposite those sides are unequal, and the greater angle is opposite the greater side.
>
> **Theorem 32:** If two angles of a triangle are unequal, then the measures of the sides opposite those angles are unequal, and the longer side is opposite the greater angle.

Yet a third inequality theorem, often referred to as "The Triangle Inequality Theorem" governs the size limit of the third side of any triangle.

> **Theorem 33:** The sum of the lengths of any two sides of a triangle is greater than the length of the third side.

Example Problems

These problems show the answers and solutions.

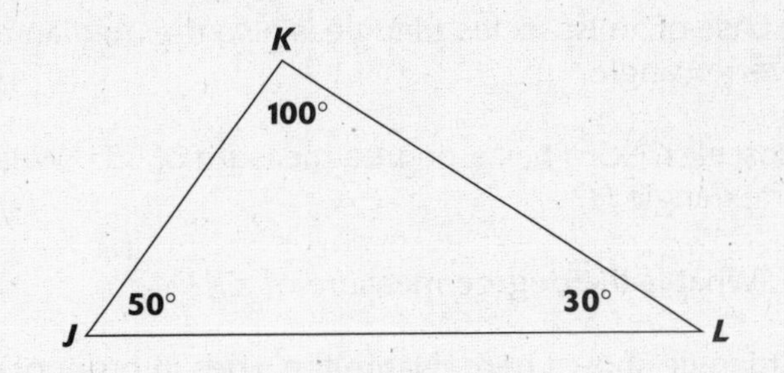

1. List the sides of $\triangle JKL$ in order of increasing length.

***Answer:* JK, KL, JL** According to Theorem 32, the sides in increasing order pair up with the angles in increasing order. The smallest angle is $\angle L$, whose opposite side is $\overline{JK}$, followed by $\angle J$, whose opposite side is $\overline{KL}$, followed by $\angle K$, whose opposite side is $\overline{JL}$.

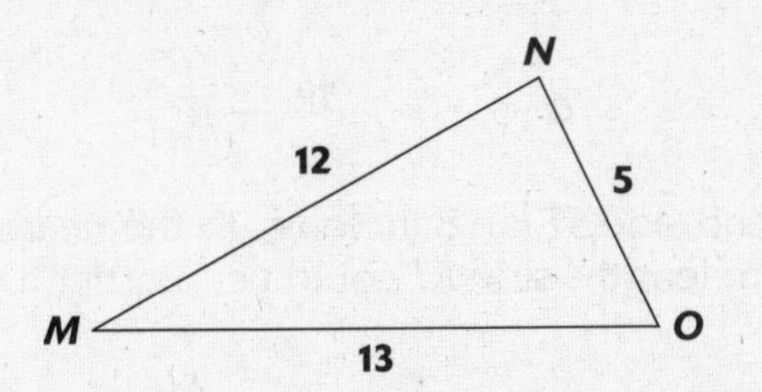

2. List the angles of $\triangle MNO$ in order of increasing angle measure.

***Answer:* ∠M, ∠O, ∠N** According to Theorem 31, the angles in increasing order pair up with the sides in increasing order. The smallest side is $\overline{NO}$, whose opposite angle is $\angle M$; followed by $\overline{MN}$, whose opposite angle is $\angle O$; followed by $\overline{MO}$, whose opposite angle is $\angle N$.

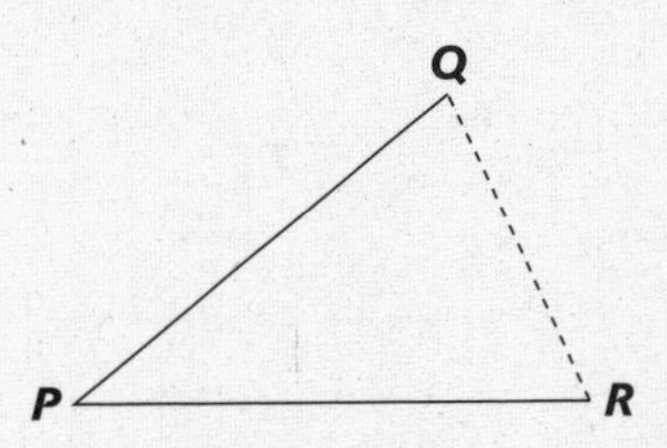

3. Side *RP* is 10 cm long, and side *PQ* is 12 cm long. To the nearest centimeter, what are the minimum and maximum lengths side $\overline{QR}$ could be in order to form $\triangle PQR$?

***Answer:* minimum 3 cm; maximum 21 cm** According to Theorem 33:

$QR + 10 \text{ cm} > 12 \text{ cm}$	**and**	$10 \text{ cm} + 12 \text{ cm} > QR$
$QR > 12 \text{ cm} - 10 \text{ cm}$		$22 \text{ cm} > QR$
$QR > 2 \text{ cm}$		$QR < 22 \text{ cm}$
$QR = 3 \text{ cm}$		$QR = 21 \text{ cm}$

Work Problems

Use these problems to give yourself additional practice.

1. The altitude to the base of an isosceles triangle is also the median to the base. Prove that it also bisects the vertex angle.

2. Base angle *G*, of isosceles $\triangle GHI$ has a degree measure of 55°. What is the degree measure of the vertex angle *H*?

3. $\triangle JKL$ is equilateral. What is the degree measure of $\angle K$?

4. Examine the right triangle shown here. Name the sides in order of decreasing length.

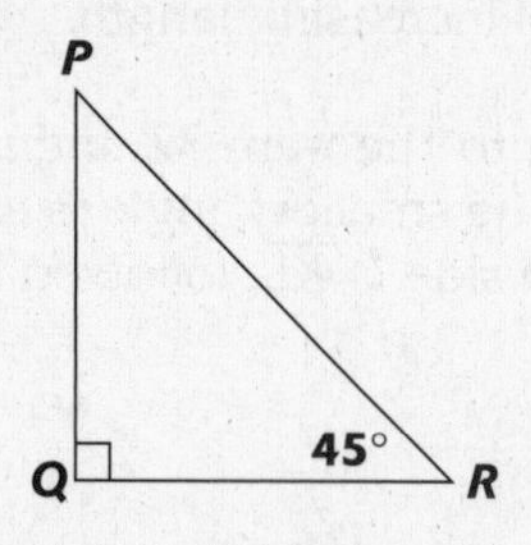

5. Side *RS* is 15 in. long, and side *ST* is 18 in. long. To the nearest inch, what are the minimum and maximum lengths side *RT* could be in order to form $\triangle RST$?

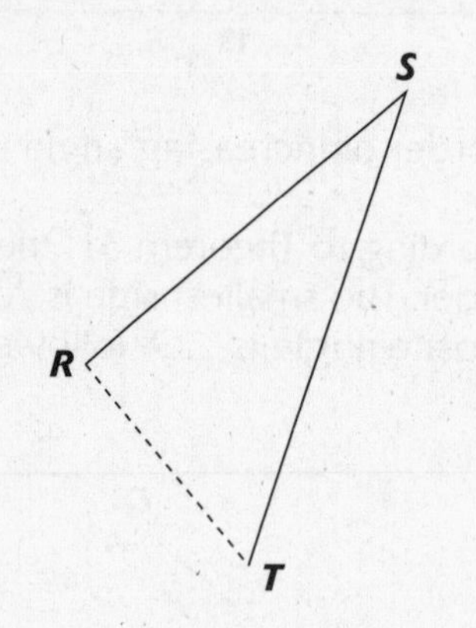

Worked Solutions

1. First, we must draw a diagram. You have heard that a picture is worth a thousand words?! Make sure to mark the information that you know.

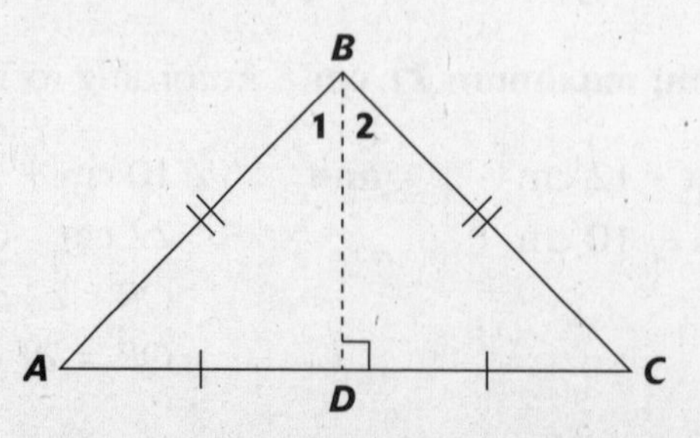

We'll use SSS, but feel free to use SAS to prove the triangles congruent.

Given: Isosceles $\triangle ABC$ with altitude/median $\overline{BD}$, as marked in the preceding diagram.

Prove: $\overline{BD}$ bisects $\angle B$.

Statement	***Reason***
1a. $\triangle ABC$ is isosceles.	1a. Given
1. $\overline{AB} \cong \overline{BC}$	1. Definition of an isosceles triangle.
2. $\overline{BD}$ is the Median to $\overline{AC}$.	2. Given
3. $\overline{AD} \cong \overline{CD}$	3. Definition of a median.
4. $\overline{BD} \cong \overline{BD}$	4. Common (or Identity).
5. $\triangle ABD \cong \triangle CBD$	5. SSS
6. $\angle 1 \cong \angle 2$	6. CPCTC
7. $\overline{BD}$ bisects $\angle B$.	7. Definition of an angle bisector.

2. **70°** If one base angle has a degree measure of 55°, the other must also, since the base angles of an isosceles triangle are the angles opposite the equal legs, and, therefore, are equal in measure. Since there are 180° total angle measure in the triangle,

$$m\angle H = 180° - (55° + 55°)$$
$$m\angle H = 180° - 110°$$
$$m\angle H = 70°$$

3. **60°** An equilateral triangle is also equiangular. That means all angles have the same measure:

$$m\angle K = 180° \div 3$$
$$m\angle K = 60°$$

4. **45°** *PR* is the longest. *QP* and *RQ* are identical in length. Since one acute angle of right $\triangle PQR$ is 45° in angle measure, the other must be the same:

$$m\angle P = 180° - (90° + 45°)$$
$$m\angle P = 180° - 135°$$
$$m\angle P = 45°$$

By Theorem 28: If two angles of a triangle are equal, then the sides opposite those angles are equal. What we have here is an isosceles right triangle.

5. **Minimum 4 in., maximum 32 in.** According to Theorem 33:

$RT + 15 \text{ in.} > 18 \text{ in.}$	and	$15 \text{ in.} + 18 \text{ in.} > RT$
$RT > 18 \text{ in.} - 15 \text{ in.}$		$33 \text{ in.} > RT$
$RT > 3 \text{ in.}$		$RT < 33 \text{ in.}$
$RT = 4 \text{ in.}$		$RT = 32 \text{ in.}$

Chapter 4

Polygons

A closed plane figure composed of line segments with three or more sides and angles is called a **polygon**, from the Greek for many (*poly*) angles (*gon*). Closed, in this definition, is for the moment an undefined term, which we'll define shortly. Just as many types of triangles exist, so do many types of polygons. We'll look at many of them, but pay special attention to the many types of quadrilaterals (four-sided polygons). We'll also study interior and exterior angle sums and medians.

Types of Polygons

The following figures are *not* polygons. The reason why they are not is written near each figure.

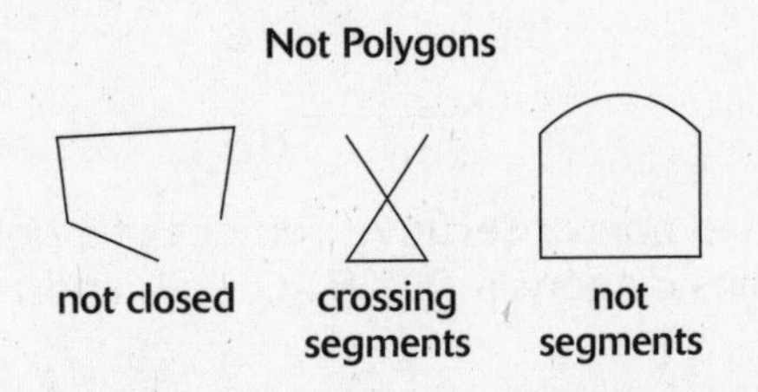

In a **closed** figure, one cannot cross from inside to outside without crossing a side. One can go from inside to outside in the first figure by passing through the opening. That's why it is not a polygon.

Polygons may be classified into two general types, **concave** and **convex**.

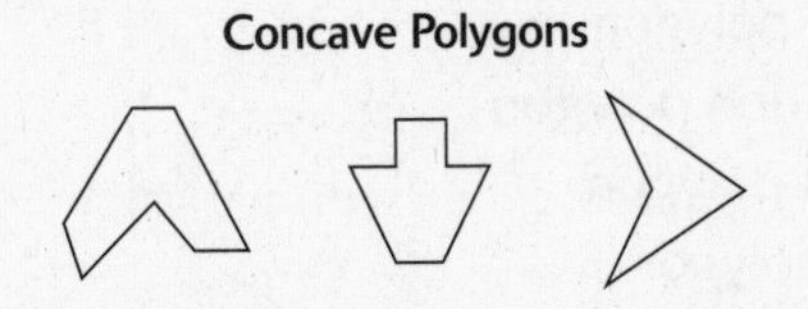

These are concave polygons. Each contains at least one interior angle greater than 180°. The middle polygon in this group contains two 270° angles.

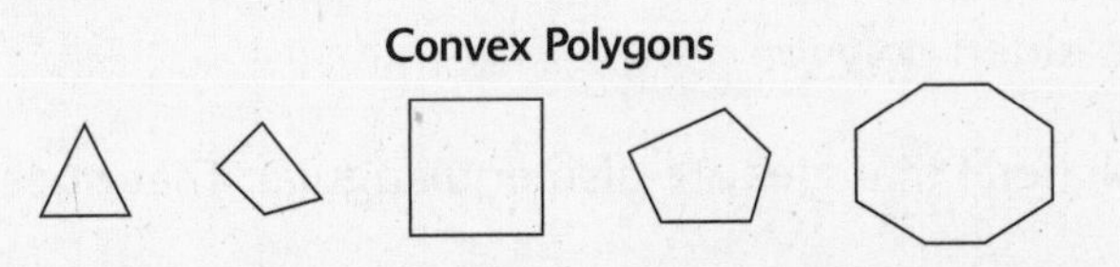

Convex polygons, shown here, contain sides that meet at angles of 180° or less. Another way to tell a concave from a convex polygon is to consider going around its outside. As you go around the outside of a convex polygon in a clockwise direction, each time you come to the end of a line segment you turn right. On a concave polygon, moving around the outside clockwise, at some point you're going to have to make a left turn.

As a rule, convex polygons are the main ones with which plane geometry is concerned.

Naming Polygons' Parts

The endpoints of each side of a polygon are called **vertices** (plural of vertex). You've seen them before in triangles. We name a polygon by assigning a letter to each vertex, proceeding in alphabetical order in a clockwise or a counterclockwise direction.

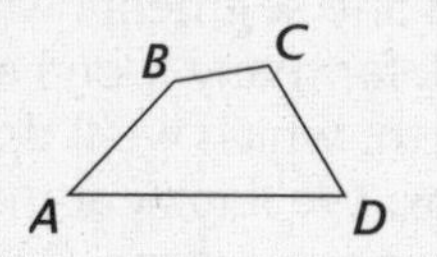

The vertices of this four-sided polygon are A, B, C, and D. Any two sides that share a common vertex are called **consecutive sides**. This polygon has four pairs of consecutive sides. They are $\overline{AB}$ and $\overline{BC}$, $\overline{BC}$ and $\overline{CD}$, $\overline{CD}$ and $\overline{AD}$, and $\overline{AD}$ and $\overline{AB}$.

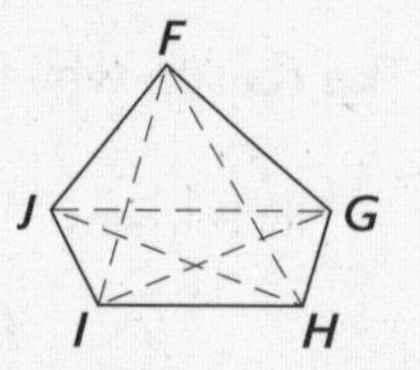

A line segment that joins any two nonconsecutive vertices of a polygon is a **diagonal**. Here, the five-sided polygon $FGHIJ$ contains diagonals $\overline{FH}$, $\overline{FI}$, $\overline{GI}$, $\overline{GJ}$, and $\overline{HJ}$.

Number of Sides and Angles

Polygons are classified into types by the number of sides and angles they have. It doesn't matter which you count, because even though polygon means many angled, the figures are also many sided—the same number as angles. The most common polygons are named:

triangle – a three-sided polygon
quadrilateral – a four-sided polygon
pentagon – a five-sided polygon
hexagon – a six-sided polygon
heptagon (sometimes septagon) – a seven-sided polygon
octagon – a eight-sided polygon
nonagon – a nine-sided polygon
decagon – a ten-sided polygon

We saw earlier that equilateral triangles are also equiangular. That does not hold true for all polygons.

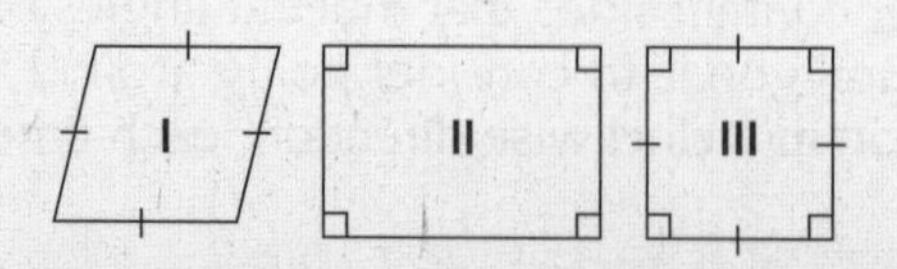

Quadrilateral I shown here is equilateral, but not equiangular. Quadrilateral II is not equilateral, but is equiangular. Quadrilateral III is both equilateral and equiangular.

Regular Polygons

When a polygon is both equiangular and equilateral, it is said to be **regular**.

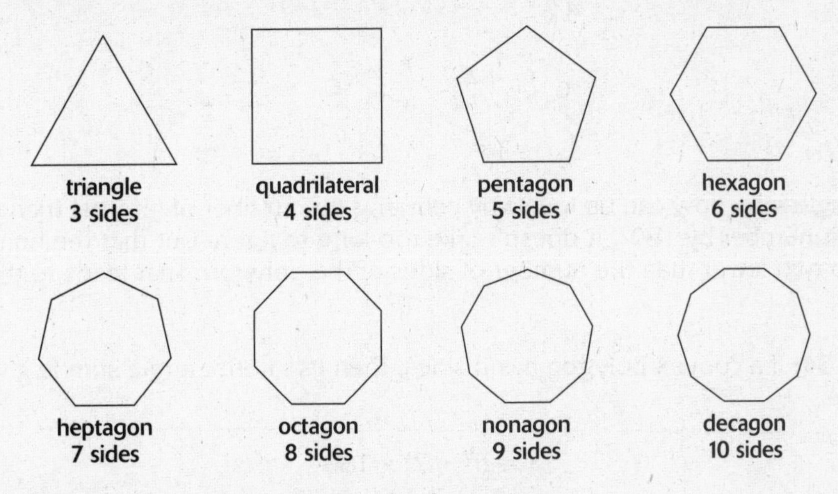

In order for a polygon to be regular, it must also be convex.

Example Problems

These problems show the answers and solutions. Refer to the following figures to solve problems 1 through 3.

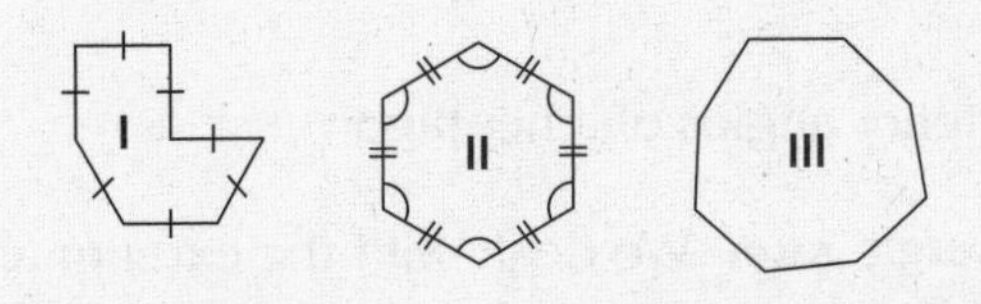

1. Tell as much information as you can about polygon I.

   ***Answer:* concave equilateral heptagon** A seven-sided figure is a heptagon. The sides are marked as equal, so it is equilateral. Because it is concave, it cannot be regular.

2. Tell as much information as you can about polygon II.

   ***Answer:* regular hexagon** A six-sided figure is a hexagon. The sides are marked as equal, so it is equilateral. Because its angles are marked as equal, it is regular, which includes the equilateral property.

3. Tell as much information as you can about polygon III.

   ***Answer:* convex octagon** An eight-sided figure is an octagon, and no angle is greater than 180°, so it's convex. Nothing is known about the equality of sides or angles.

Angle Sums

If we start out with a polygon of four or more sides, drawing all the diagonals possible from a single vertex will produce a series of triangles that do not overlap, as already noted in an earlier chapter.

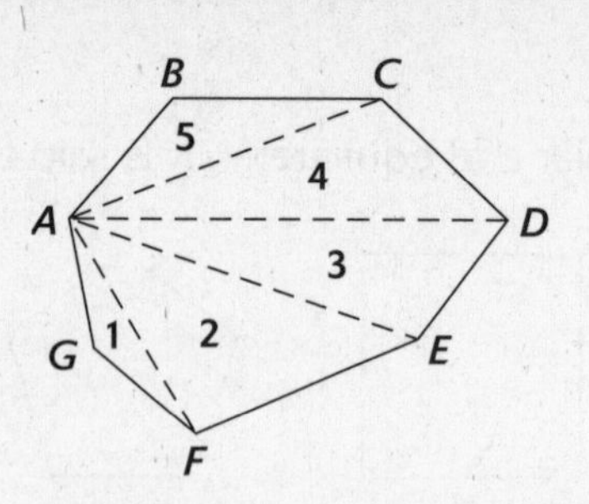

The **interior angle sum** now can be found by counting the number of internal triangles and then multiplying that number by 180°. It doesn't take too long to figure out that the number of triangles is equal to two fewer than the number of sides of the polygon. That leads to the following theorem:

Theorem 34: If a convex polygon has n sides, then its interior angle sum is given by the equation:

$$S = (n - 2) \times 180°$$

Since the preceding polygon has seven sides, what we found by observation, we can now find by formula:

$S = (7 - 2)$ nonoverlapping triangles $\times\ 180°$
$S = 5 \times 180°$
$S = 900°$

That's the sum of the interior angles of a heptagon.

Now let's consider **exterior angle sum**. We dealt with the exterior angles of a triangle in the last chapter, but not with their sum.

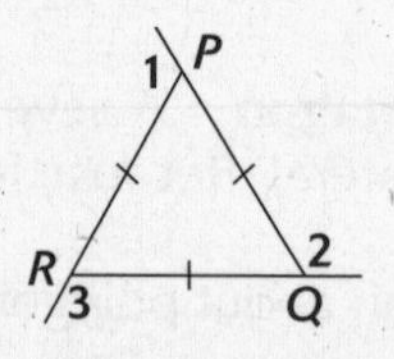

Since $\triangle PQR$ is an equilateral triangle, it is also equilangular, and each of the interior angles has a degree measure of 60°. This means that each of the external angles marked, $\angle 1$, $\angle 2$, and $\angle 3$, must have a degree measure of 120°, since each is supplementary to a 60° angle. This makes the following the sum of the exterior angles:

$$120° + 120° + 120° = 360°$$

That leads to a rather startling conclusion that is expressed as the following theorem.

Theorem 35: If a polygon is convex, then the sum of the exterior angles (drawn one per vertex) is 360°.

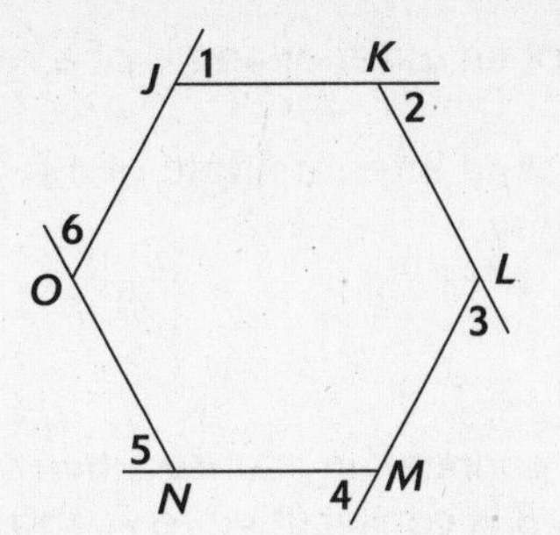

Here, we have a regular hexagon with an exterior angle at each vertex. We can find the measure of each of those exterior angles by dividing 360° by the number of vertices, 6. $360° \div 6 = 60°$. If each exterior angle of a regular hexagon is 60°, then the degree measure of each interior angle must be the supplement of 60°, since each exterior angle is supplementary to each interior angle in any regular polygon. This means the degree measure of each interior angle of a regular hexagon is $180° - 60° = 120°$.

Example Problems

These problems show the answers and solutions.

1. What's the sum of the exterior angles of a regular nonagon (drawn one per vertex)?

***Answer:* 360°** Remember Theorem 35: If a polygon is convex, then the sum of the exterior angles (drawn one per vertex) is 360°.

2. What's the degree measure of an exterior angle of a regular nonagon?

***Answer:* 40°** A nonagon has nine vertices, so divide the measure of the sum of the exterior angles by 9:

$$360° \div 9 = 40°$$

3. What's the degree measure of an interior angle of a regular nonagon?

***Answer:* 140°** Since each exterior angle is supplementary to each interior angle in any regular polygon, the degree measure of each interior angle of a regular nonagon is

$$180° - 40° = 140°$$

Work Problems

Use these problems to give yourself additional practice.

1. Name all diagonals of convex polygon *ABCD*.

2. Name all diagonals of convex polygon *GHIJK*.

3. What's the sum of the exterior angles of a regular decagon (drawn one per vertex)?

4. What's the degree measure of an exterior angle of a regular decagon?

5. What's the degree measure of an interior angle of a regular decagon?

Worked Solutions

1. **$\overline{AC}$, $\overline{BD}$** A diagonal cannot connect any consecutive vertices. *A* is consecutive to *B* and *D*, and so can connect to only *C*. *B* is consecutive to *A* and *C*, and so forms a diagonal with *D*.

2. **$\overline{GI}$, $\overline{GJ}$, $\overline{HJ}$, $\overline{HK}$, $\overline{IK}$** Start with the first vertex, *G*, and skip *H* to form $\overline{GI}$ and $\overline{GJ}$. Move on to the second vertex, *H*, and skip *I* to form $\overline{HJ}$ and $\overline{HK}$. Move on to *I* and skip *J* to form $\overline{IK}$. That's it, since any other combination would coincide with an existing diagonal. Note that each vertex was used twice.

3. **360°** Remember Theorem 35: If a polygon is convex, then the sum of the exterior angles (drawn one per vertex) is 360°.

4. **36°** A decagon has 10 vertices, so divide the measure of the sum of the exterior angles by 10:

$$360° \div 10 = 36°$$

5. **144°** Since each exterior angle is supplementary to each interior angle in any regular polygon, the degree measure of each interior angle of a regular nonagon is

$$180° - 36° = 144°$$

Quadrilaterals

Many geometry books devote a full chapter or more to quadrilaterals because there are many different types of quadrilateral, and they are not all created equal.

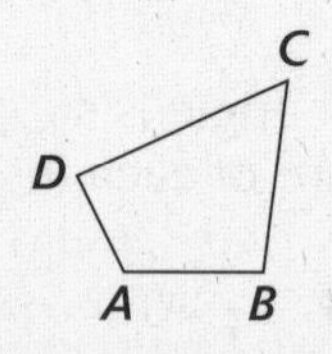

ABCD, in the preceding figure, is a run-of-the-mill quadrilateral with none of its four sides parallel or equal and all angles different. By Theorem 34, we can readily show that the sum of its interior angle measure is 360°. Now, let's fancy up the quadrilateral by making a pair of its sides parallel.

Trapezoids

A quadrilateral with one pair of opposite sides parallel is called a **trapezoid**. The parallel sides are called the **bases**, and the *non*parallel sides are the **legs**.

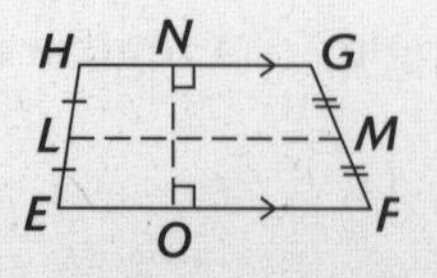

Any segment that is perpendicular to both bases is called an **altitude** of the trapezoid. The length of that segment is the trapezoid's **height**. A segment joining the midpoints of the trapezoid's legs is known as its **median**.

$\overline{GH}$ and $\overline{EF}$ are bases of trapezoid *EFGH*.
$\overline{EH}$ and $\overline{FG}$ are its legs.
$\overline{LM}$ is the median of trapezoid *EFGH*.
NO, the length of the altitude $\overline{NO}$, is its height.

Now suppose that we were to make the other sides of the trapezoid parallel . . .

Parallelograms

A quadrilateral with both pairs of opposite sides parallel is called a **parallelogram**. Each pair of parallel sides is called a **pair of bases** of the parallelogram. Any segment perpendicular to either pair of bases is an altitude, whose length is the height of the parallelogram.

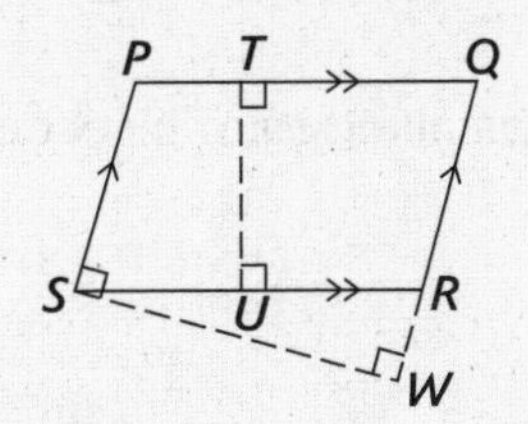

The preceding figure shows both sets of bases and altitudes. The symbol ▱ is used as mathematical shorthand for the word parallelogram.

In ▱*PQRS*

$\overline{TU}$ is an altitude to bases $\overline{PQ}$ and $\overline{RS}$.
$\overline{SW}$ is an altitude to bases $\overline{PS}$ and $\overline{QR}$.
$\overline{TU}$ is the height of ▱*PQRS* with $\overline{PQ}$ and $\overline{RS}$ as bases. *SW* is the height of ▱*PQRS* with $\overline{PS}$ and $\overline{QR}$ as bases.

Here are some theorems that pertain to parallelograms:

Theorem 36: Either diagonal of a parallelogram divides it into two congruent triangles.

Here are two congruent triangles created by the diagonal of a parallelogram.

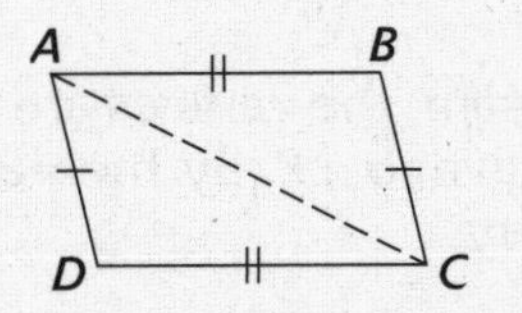

According to Theorem 36, in ▱*ABCD*, $\triangle ABC \cong \triangle CDA$. That leads to three very easily proven theorems:

Theorem 37: Opposite sides of a parallelogram are congruent.
Theorem 38: Opposite angles of a parallelogram are congruent.
Theorem 39: Consecutive angles of a parallelogram are supplementary.

In ▱$ABCD$

By Theorem 37, $\overline{AB} \cong \overline{CD}$ and $\overline{AD} \cong \overline{BC}$.
By Theorem 38, $m\angle A \cong m\angle C$ and $m\angle D \cong m\angle B$.
By Theorem 39:

$\angle A$ and $\angle B$ are supplementary.
$\angle B$ and $\angle C$ are supplementary.
$\angle C$ and $\angle D$ are supplementary.
$\angle D$ and $\angle A$ are supplementary.

Now, we have Theorem 40:

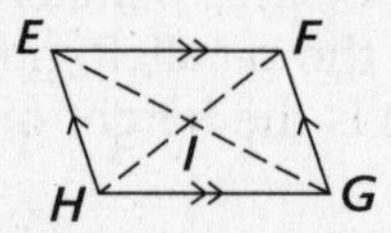

Theorem 40: The diagonals of a parallelogram bisect each other.

In ▱$EFGH$, $EI = GI$ and $FI = HI$.

Example Problems

These problems show the answers and solutions.

1. What type of polygon contains exactly two bases and exactly two legs?

   ***Answer:* a trapezoid** With exactly two bases and exactly two legs, the figure must be a quadrilateral. Two bases in terms of a quadrilateral means two parallel sides. That's the definition of a trapezoid.

2. In ▱$ABCD$, $m\angle B = 130°$. Find $m\angle D$.

   ***Answer:* 130°** In a parallelogram, the vertices are labeled consecutively. That means that $\angle D$ is going to be opposite $\angle B$. By Theorem 38, opposite angles of a parallelogram are congruent.

3. In ▱$PQRS$, $m\angle S = 70°$. Find $m\angle R$.

   ***Answer:* 110°** In a parallelogram, the vertices are labeled consecutively. That means that $\angle S$ is going to be consecutive to $\angle R$. By Theorem 39, consecutive angles of a parallelogram are supplementary.

$$180° - 70° = 110°$$

Proofs of Parallelograms

Often, it is necessary to prove that a quadrilateral is a parallelogram. The following theorems describe tests that you may use to determine whether or not a figure in question is a parallelogram.

Theorem 41: If both pairs of opposite sides of a quadrilateral are equal, then it is a parallelogram.

Theorem 42: If both pairs of opposite angles of a quadrilateral are equal, then it is a parallelogram.

Theorem 43: If all pairs of consecutive angles of a quadrilateral are supplementary, then it is a parallelogram.

Theorem 44: If one pair of opposite sides of a quadrilateral are both equal and parallel, then it is a parallelogram.

Theorem 45: If the diagonals of a quadrilateral bisect each other, then it is a parallelogram.

Example Problems

These problems show the answers and solutions.

1. Prove that quadrilateral *ABCD*, shown in the following figure, is a parallelogram.

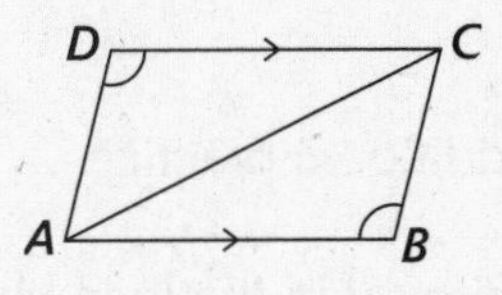

***Answer:* Proof**

Statement	***Reason***
1. $\angle D = \angle B$	1. Given
2. $\overline{AC} = \overline{AC}$	2. Identity
3. $\overline{CD} \parallel \overline{AB}$	3. Given
4. $m\angle DCA = m\angle BAC$	4. Alt. interior angles of parallel lines
5. $\triangle ACD \cong \triangle CAB$	5. SAA
6. $\overline{AD} \cong \overline{BC}$	6. CPCTC
7. $\overline{AB} \cong \overline{CD}$	7. CPCTC
8. Quadrilateral *ABCD* is a ▱.	8. Theorem 41

Please note that reason #8, Theorem 41, is just a shortcut to avoid having to write "If both pairs of opposite sides of a quadrilateral are equal, then it is a parallelogram." We use it to conserve space. Feel free to write out the actual reason if you desire.

2. Prove that quadrilateral *LMNO*, shown in the following figure, is a parallelogram.

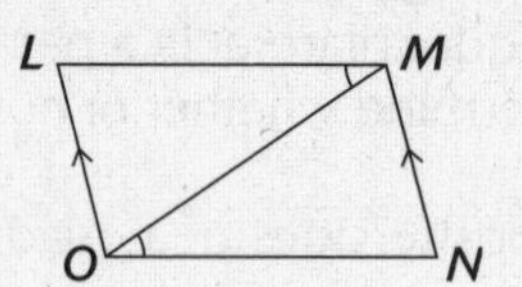

***Answer:* Proof**

Statement	*Reason*
1. $\angle LMO \cong \angle MON$	1. Given
2. $\overline{LO} \parallel \overline{MN}$	2. Given
3. $\angle NMO \cong \angle LOM$	3. Alt. interior angles of parallel lines
4. $\triangle LMO \cong \triangle NOM$	4. ASA
5. $\overline{LO} \cong \overline{MN}$	5. CPCTC
6. Quadrilateral *LMNO* is a parallelogram.	6. Theorem 44

Work Problems

Use these problems to give yourself additional practice.

1. Prove that quadrilateral *WXYZ*, shown in the following figure, is a parallelogram because of bisecting diagonals (Theorem 45).

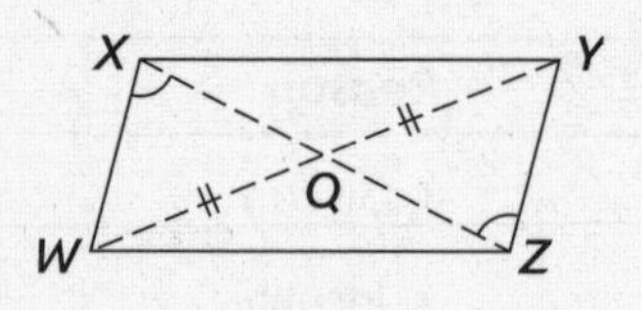

Use the figure that follows for problems 2 through 4.

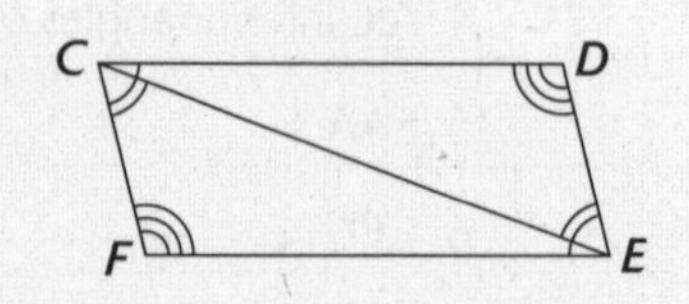

2. Prove that quadrilateral *CDEF* is a parallelogram by both pairs of opposite angles being equal (Theorem 42).

3. Prove that quadrilateral *CDEF* is a parallelogram by both pairs of opposite sides being equal (Theorem 41).

4. Prove that $\angle C$ is supplementary to $\angle D$ and to $\angle F$.

5. Prove that quadrilateral *JKLM*, in the following figure, is a parallelogram because one pair of opposite sides are both equal and parallel (Theorem 44).

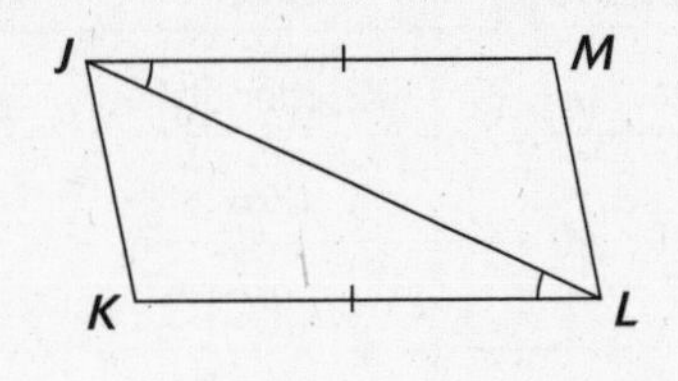

Worked Solutions

1. Proof follows.

Statement	*Reason*
1. $\overline{QW} = \overline{QY}$	1. Given
2. $m\angle WXQ = m\angle QZY$	2. Given
3. $m\angle WQX = m\angle ZQY$	3. Vertical angles are equal.
4. $\triangle WQX \cong \triangle YQZ$	4. SAA
5. $\overline{QX} \cong \overline{QZ}$, $\overline{XZ} = \overline{XQ} + \overline{QZ}$	5. CPCTC
6. $\overline{WY} = \overline{WQ} + \overline{QY}$	6. Segment addition postulate (Postulate 8)
7. $\overline{WY}$ and $\overline{XZ}$ bisect each other.	7. Definition of bisect
8. Quadrilateral *WXYZ* is a ▱.	8. Theorem 45

2. Proof follows.

Statement	*Reason*
1. $m\angle D = m\angle F$	1. Given
2. $m\angle DCE = m\angle CEF$	2. Given
3. $m\angle ECF = m\angle CED$	3. Given
4. $m\angle DCF = m\angle DEF$	4. Angle addition postulate (Postulate 10), and when equals are added to equals the results are equal.
5. Quadrilateral *CDEF* is a ▱.	5. Theorem 42

3. Proof follows.

Statement	***Reason***
1. Steps 2 and 3 from Problem 2, above.	1. Given
2. $\overline{CE} = \overline{CE}$	2. Identity
3. $\triangle CDE \cong \triangle EFC$	3. ASA
4. $\overline{CD} \cong \overline{FE}$	4. CPCTC
5. $\overline{CF} \cong \overline{DE}$	5. CPCTC
6. Quadrilateral *CDEF* is a ▱.	6. Theorem 41

4. Proof Begin with the last line of either of the last two solutions, having proven Quadrilateral *CDEF* is a ▱.

Statement	***Reason***
1. $\angle C$ is supplementary to $\angle D$	1. Theorem 39
2. $\angle C$ is supplementary to $\angle F$	2. Theorem 39

5. Proof follows.

Statement	***Reason***
1. $m\angle MJL = m\angle KLJ$	1. Given
2. $\overline{MJ} \parallel \overline{KL}$	2. If Alt. Int. $\angle$s =, the lines are parallel (Post. 12).
3. $\overline{MJ} = \overline{KL}$	3. Given
4. Quadrilateral *JKLM* is a ▱.	4. Theorem 44

Special Parallelograms

Just as all quadrilaterals are not created equal, neither are all parallelograms. Indeed, some are very special and possess very special properties. Just bear in mind as you consider each of these special cases, that each is indeed a parallelogram, and all the properties of parallelograms that we've studied apply to them, in addition to what sets them apart.

Rectangle

Add a single right angle to a parallelogram, and it becomes a **rectangle**. Only one right angle is needed, since consecutive angles of a parallelogram are supplementary (remember Theorem 39?). Think about it.

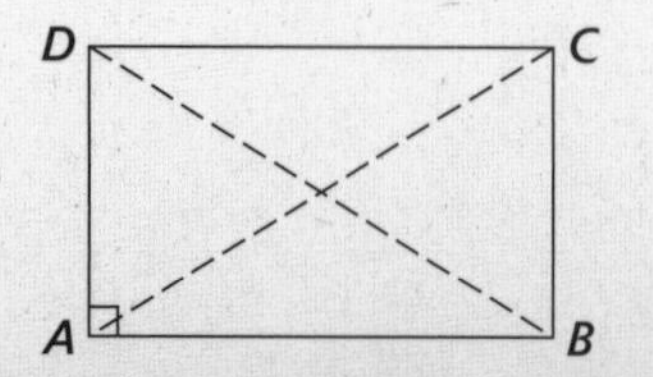

If you look at triangles ACB and BDA, you'll see that they are congruent by LL. That means $\overline{BD}$ is $\cong$ to $\overline{AC}$ by CPCTC, which leads to the following theorem.

Theorem 46: The diagonals of a rectangle are equal.

Rhombus

A parallelogram that is equilateral is known as a **rhombus**.

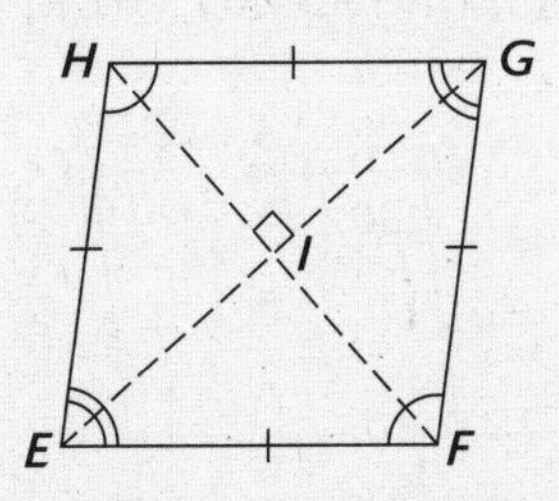

A rhombus has two additional properties not found in other parallelograms.

Theorem 47: The diagonals of a rhombus bisect the opposite angles.
Theorem 48: The diagonals of a rhombus are perpendicular to one another.

The preceding diagram illustrates both Theorems 47 and 48.

Square

An equilateral parallelogram with one right angle is a **square**.

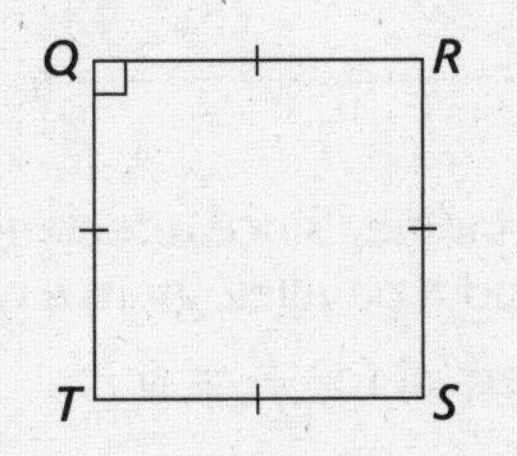

Hopefully, the diagram that follows will help you to understand why a square has the properties it does.

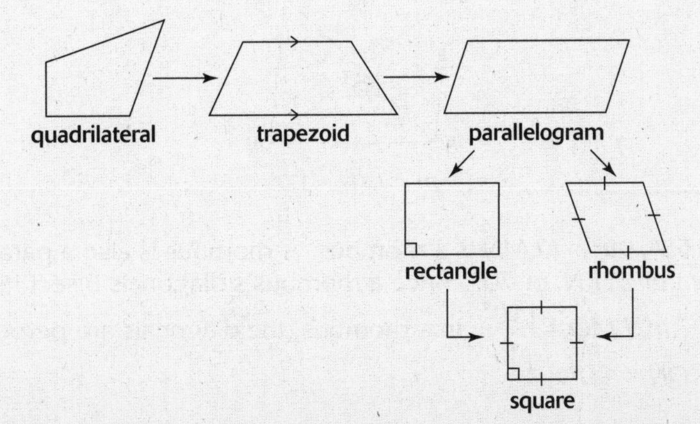

You can get to a square by making a rectangle equilateral or by adding a right angle to a rhombus. A square is a parallelogram, a rectangle, and a rhombus, and so has all the properties of all of those.

Example Problems

These problems show the answers and solutions.

1. The following figure is as marked. Find RS, RU, $m\angle R$, $m\angle S$, $m\angle T$

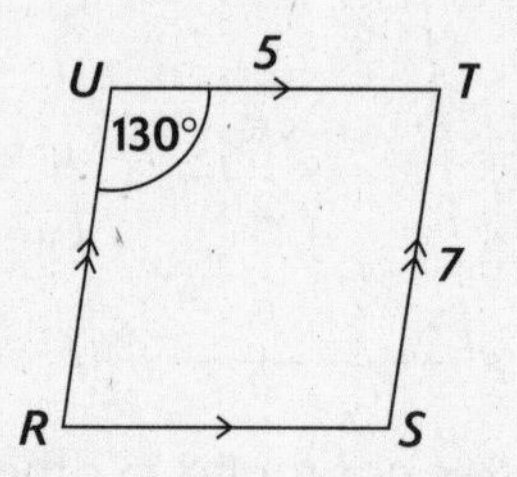

Answer: 5, 7, 50°, 130°, 50° *RSTU* is a parallelogram. In a parallelogram, the opposite sides are equal, so $RS = UT$ (5), and $RU = ST$ (7). Consecutive angles are supplementary, so $m\angle R = 180° - 130°$. Additionally, opposite angles are congruent, so $m\angle S = m\angle U$ (130°) and $m\angle T = m\angle R$ (50°).

2. The following figure is as marked. Find DF, CG, and DG.

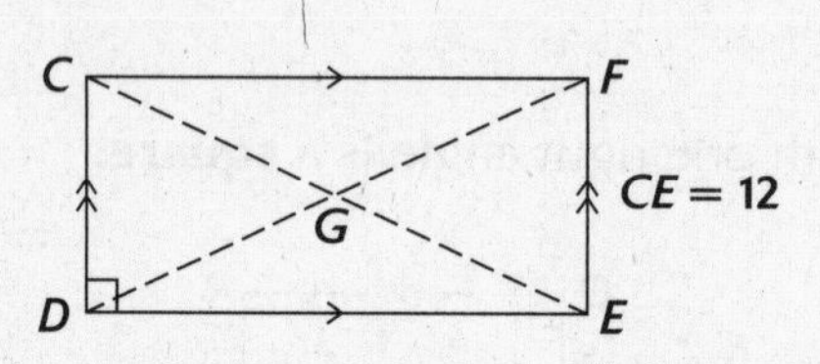

Answer: 12, 6, 6 *CDEF* is a rectangle. Since a rectangle's diagonals are equal, $DF = CE$ (12). A rectangle is a parallelogram, and a parallelogram's diagonals bisect each other. That makes each half diagonal (CG and DG) equal to $\frac{12}{2} = 6$.

3. The figure that follows is as marked. Find $m\angle KLO$, $m\angle LMO$, and $m\angle KON$.

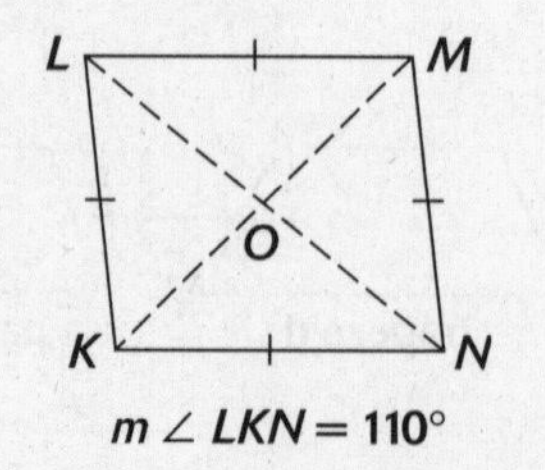

$m\angle LKN = 110°$

Answer: 35°, 55°, 90° *KLMN* is a rhombus. A rhombus is also a parallelogram, so $\angle L$ is the supplement of $\angle LKN$, or 70°. Since a rhombus's diagonals bisect its opposite angles, $m\angle KLO = \frac{70°}{2}$, $m\angle LMO = \frac{110°}{2}$. In a rhombus, the diagonals are perpendicular to each other, so $m\angle KON = 90°$.

Special Trapezoids

Special trapezoids also exist; the following figure shows one of these special shapes.

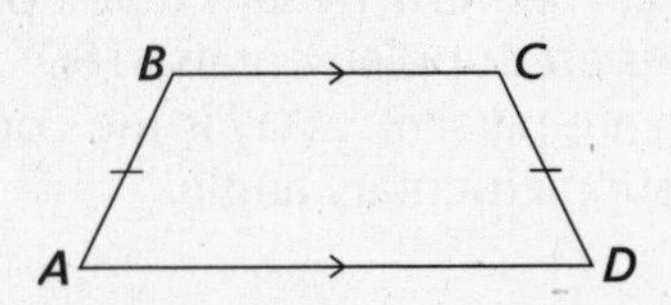

Did you notice that the two legs of this trapezoid are marked as being equal in length? That makes it an **isosceles trapezoid**. Any pair of angles that share a single base of a trapezoid are referred to as the trapezoid's **base angles**. So, unlike an isosceles triangle, an isosceles trapezoid has two pairs of base angles; in this case, $\angle A$ and $\angle D$, and $\angle B$ and $\angle C$. That gives us two new theorems:

Theorem 49: Base angles of an isosceles trapezoid are equal.

That means that in trapezoid *ABCD*, $m\angle A = m\angle D$, and $m\angle B = m\angle C$.

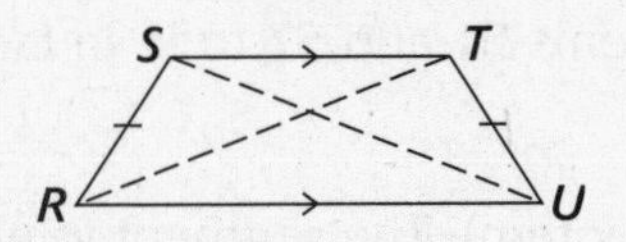

Theorem 50: Diagonals of an isosceles trapezoid are equal.

By Theorem 50, $SU = RT$.

You may remember that the median of a trapezoid is the line segment that joins the midpoints of the legs, like *XY* in the following figure.

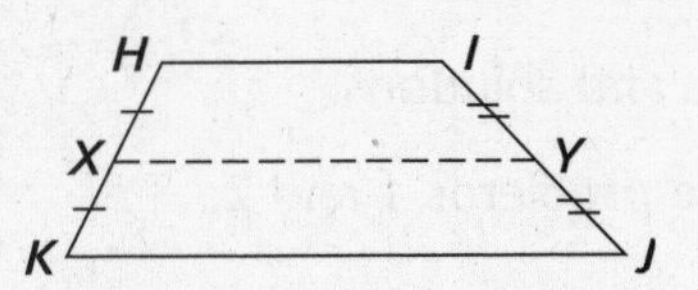

Theorem 51: The median of any trapezoid is parallel to both bases.
Theorem 52: The median of any trapezoid is half the length of the sum of the bases.

Example Problems

These problems show the answers and solutions.

Use trapezoid *HIJK* to answer questions 1 and 2.

1. Suppose $HI = 15$ and $JK = 25$. Find the length of *XY*.

   ***Answer:* 20** By Theorem 52, the median of any trapezoid is half the length of the sum of the bases.

   $XY = \frac{1}{2}(15 + 25)$ $\quad$ $XY = \frac{1}{2}(40)$ $\quad$ $XY = 20$

2. Suppose that $m\angle I = 130°$. Find $m\angle IYX$.

***Answer:* 50°** According to Theorem 51, the median of any trapezoid is parallel to both bases. Going back to Chapter 2, Theorem 14 says that if parallel lines are cut by a transversal, their consecutive interior angles are supplementary. $180° - 130° = 50°$. Failing to recall that, since the two lines in question are parallel, $\angle XYJ$ is the corresponding angle to $\angle I$, and so is 130°. $\angle IYX$ is its consecutive supplementary angle.

The Midpoint Theorem

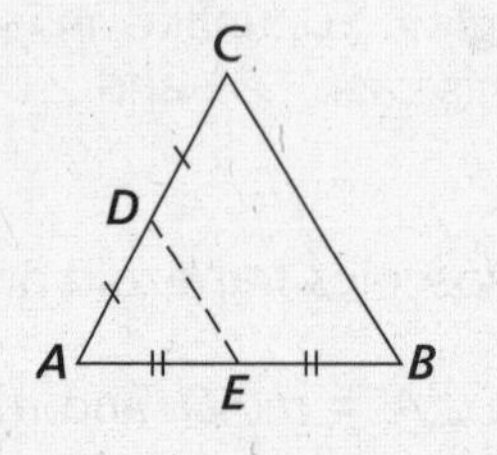

If you look at $\triangle ABC$ as a trapezoid with one base $\overline{BC}$ and its other base so small as to be zero in length, at A, you could apply Theorems 51 and 52 to it. In fact, we have such a theorem especially for triangles:

> **Theorem 53** (The Midpoint Theorem): The segment joining the midpoints of any two sides of a triangle is half the length of the third side and parallel to it.

The reason segment $\overline{DE}$ is half the third side is that the sum of the third side and 0 is the length of the third side. Then we halve that sum, just as we did for the trapezoid in Theorem 51.

Example Problems

These problems show the answers and solutions.

Use the figure that follows to solve problems 1 and 2.

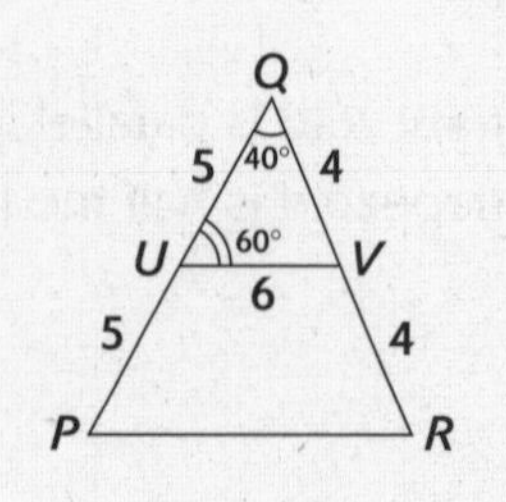

1. Find PR.

***Answer:* 12** According to the Midpoint Theorem, the segment joining the midpoints of any two sides of a triangle is half the length of the third side. That makes the third side twice the length of UV: $2 \times 6 = 12$. Or, if you prefer to work it algebraically,

$$\begin{aligned} UV &= \frac{PR}{2} \\ 2UV &= PR \\ 2(6) &= PR \\ PR &= 12 \end{aligned}$$

2. Find $m\angle R$.

***Answer:* 80°** Since $\angle Q = 40°$ and $\angle QUV = 60°$, $\angle QVU = 80°$. (A triangle's internal angles sum to 180°.)

By the Midpoint Theorem, $\overline{UV} \parallel \overline{PR}$. That makes $\angle R$ the corresponding angle to $\angle QVU$ on transversal $\overline{QR}$. Since corresponding angles of parallel lines are equal, $\angle R = 80°$.

Work Problems

Use these problems to give yourself additional practice.

1. In $\square ABCD$, $\angle C$ is a right angle. AC is 10. $AB = 5$. Find BD. Explain your answer.
2. In $\square QRST$, $QR = RS$. Diagonals $QS = RT$. What is $m\angle T$? Explain your answer.
3. Is the following statement true or false? All quadrilaterals are convex.
4. In $\square MNOP$, $\overline{MO} \perp \overline{NP}$. What conclusion can you draw?
5. In isosceles trapezoid $ABCD$, with base $\overline{AB}$, $m\angle A = 55°$. Find $m\angle C$.

Worked Solutions

1. **10** If you draw the diagram, you'll see that $\square ABCD$ is a rectangle, since one right angle is all that it takes to make a parallelogram a rectangle. Since a polygon's vertices are labeled consecutively, $\overline{AC}$ must be a diagonal of the rectangle; so must $\overline{BD}$. By Theorem 46, the diagonals of a rectangle are equal.
2. **90°** If two consecutive sides, QR and RS are equal, then $\square QRST$ is a rhombus. If diagonals $QS = RT$, then $\square QRST$ is a rectangle—actually, a square. In any event, rectangle, or square, each vertex angle must be 90°.
3. **False** The diagram that follows illustrates a concave quadrilateral.

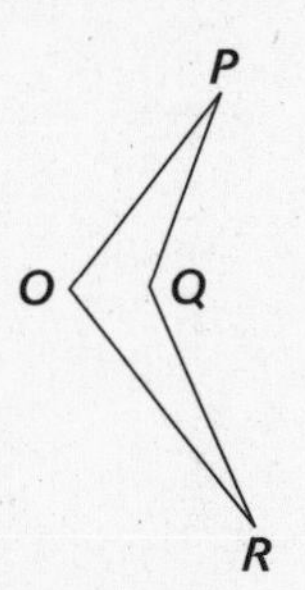

Interestingly, all triangles must be convex.

4. ***$\square$MNOP* is a rhombus.** By Theorem 48, the diagonals of a rhombus are perpendicular to one another. Its converse would be if the diagonals of a parallelogram are perpendicular to each other, then the figure is a rhombus.

5. **125°** In an isosceles trapezoid, both pairs of base angles are equal (Theorem 49). You were told one base is $\overline{AB}$ and that $m\angle A = 55°$, so $m\angle B = 55°$. That leaves $\angle$s C and D to share the remaining 250°.

$$360° - (55° + 55°) = 360° - 110°$$
$$360° - 110° = 250°$$
$$250° \div 2 = 125°$$

Chapter 5

Perimeter and Area

Perimeter is the distance around a figure. In the case of a regularly shaped fenced-in field, the perimeter of that field is the length of the fence. That length is always expressed in units of linear measure, such as inches, feet, centimeters, kilometers, and so on. **Area** is the measure of the interior of a plane figure. Picture a kitchen floor covered with squares of tile. Each tile is one foot long and one foot wide, or one square foot. Suppose that 50 of those tiles cover the kitchen's floor. Then the area of that kitchen floor is 50 square feet, which is abbreviated 50 ft^2. Area is always expressed in square units, regardless of what the shape of the surface is that's being measured; square inches, square centimeters, and square miles are abbreviated in^2, cm^2, and mi^2.

Squares and Rectangles

It is customary to use the symbol P to stand for perimeter and A to stand for area. For all polygons, the perimeter is found by adding all of the sides together. There are, however, special formulas for squares and rectangles, as well as some other figures.

Finding the Perimeter

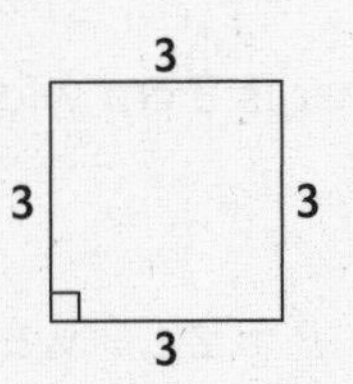

You can find the perimeter of this typical square by the formula mentioned below.

$$P_{\text{square}} = s + s + s + s$$

Of course, this formula is a waste of both time and space when dealing with a square, whose sides are all the same length. Let's take advantage of this fact and go straight to the fact that adding the same number four times can be expressed more simply as multiplication by 4. So, for a square:

$$P_{\text{square}} = 4s$$

Hopefully, you'll realize that this also applies to a rhombus—but we're getting ahead of ourselves.

In a rectangle, two of the sides are the length of the figure, and two sides are the width, so to find the perimeter, we could do the following:

$$P = l + w + l + w$$

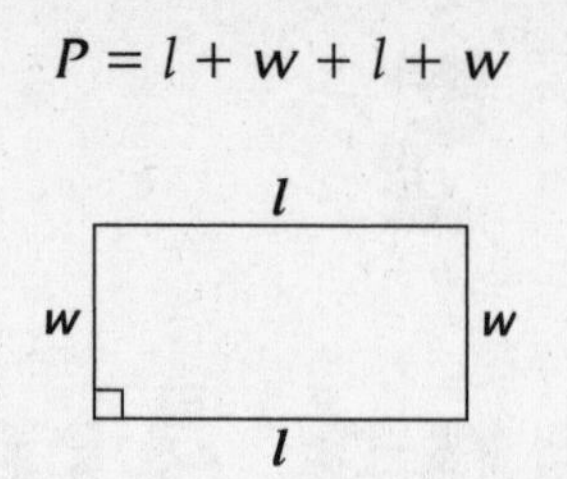

Once again, there's an awful lot of extra writing being done there. We are really adding two lengths and two widths, so that we might do either of the following:

$$P_{\text{rectangle}} = 2l + 2w$$

Or, more simply:

$$P_{\text{rectangle}} = 2(l + w)$$

Example Problems

These problems show the answers and solutions.

1. A square has a side 12 inches long. What is its perimeter?

   ***Answer:* 48 in.** For a square,

   $P = 4s$
   $P = 4(12)$
   $P = 48$ in. (The unit in the question was inches, so the answer is also in inches.)

2. Find the perimeter of a rectangle that is 8 cm wide and 13 cm long.

   ***Answer:* 42 cm**

   $P = 2(l + w)$
   $P = 2(13 + 8)$
   $P = 2(21)$
   $P = 42$ cm

Finding the Area

Area is found by multiplication. Look at the dimensions of the preceding rectangle. Now count the number of square units inside the rectangle. Can you see the formula?

$$A_{\text{rectangle}} = lw$$

To prove it, use the dimensions of the rectangle as shown:

$$A_{\text{rectangle}} = lw$$
$$A_{\text{rectangle}} = 7 \times 4$$
$$A_{\text{rectangle}} = 28 \text{ u}^2 \quad \text{(Area is always expressed in square units.)}$$

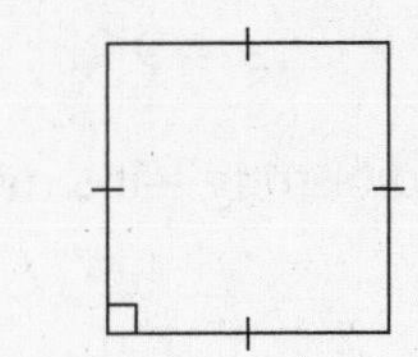

Since all sides of a square are the same length,

$A_{\text{rectangle}} = lw$ can be rewritten for a square:

$A_{\text{square}} = s \cdot s$, but this simplifies to

$A_{\text{square}} = s^2$

Suppose that the side of this square is 5 cm long. Then its area can be found like this:

$$A_{\text{square}} = s^2$$
$$A_{\text{square}} = 5^2$$
$$A_{\text{square}} = 25 \text{ cm}^2 \quad \text{Remember: Area means square units.}$$

Example Problems

These problems show the answers and solutions.

1. A square has one side 12 cm long. What is its area?

   ***Answer:* 144 cm^2**

$$A_{\text{square}} = s^2$$
$$A_{\text{square}} = 12^2$$
$$A_{\text{square}} = 144 \text{ cm}^2$$

2. A rectangle is 23 inches long and 6 inches wide. What is its area?

   ***Answer:* 138 in^2**

$$A_{\text{rectangle}} = lw$$
$$A_{\text{rectangle}} = 23 \times 6$$
$$A_{\text{rectangle}} = 138 \text{ in}^2$$

3. A rectangular playground has an area of 630 m^2. It is 10 m wide. How long is it?

   ***Answer:* 63 m**

$$A_{\text{rectangle}} = lw$$
$$630 = 10l$$
$$\frac{630}{10} = \frac{10l}{10}$$
$$l = 63 \text{ m}$$

4. A square garden is 81 ft^2. How long is its northern side?

***Answer:* 9 ft.**

$$A_{square} = s^2$$
$$81 = s^2$$
$$\sqrt{81} = \sqrt{s^2}$$
$$s = 9 \text{ ft.}$$

The "northern side" in the problem is a red herring—it's meaningless. If it is a square, then all sides are the same length.

Work Problems

Use these problems to give yourself additional practice.

1. Whoville's town square is 3 kilometers long on each side. What is the town square's perimeter?
2. A rectangular-shaped dock is 15 yards long and 7 feet wide. What is its perimeter?
3. Diamond State Forest is actually in the shape of a square. It is 100 km on each side. What is its area?
4. An area rug is 9 × 12 ft. How large an area does it cover?
5. A rectangular boardwalk is 10 m wide and has a perimeter of 240 meters. How long is it?

Worked Solutions

1. **12 km**

$$P_{square} = 4s$$
$$P_{square} = 4 \cdot 3$$
$$P_{square} = 12 \text{ km}$$

2. **104 ft.** Change yards to feet. There are 3 feet to a yard, so 15 yd = 15 · 3 ft = 45 ft.

Then:

$$P_{rectangle} = 2(l + w)$$
$$P_{rectangle} = 2(45 + 7)$$
$$P_{rectangle} = 2(52)$$
$$P_{rectangle} = 104 \text{ ft.}$$

3. **10,000 km^2**

$$A_{square} = s^2$$
$$A_{square} = 100^2$$
$$A_{square} = 10{,}000 \text{ km}^2$$

4. **108 ft²**

$$A_{\text{rectangle}} = lw$$
$$A_{\text{rectangle}} = 12 \times 9$$
$$A_{\text{rectangle}} = 108 \text{ ft}^2$$

5. **110 m**

$$P_{\text{rectangle}} = 2(l + w)$$
$$240 = 2(l + 10)$$
$$240 = 2l + 20$$
$$240 - 20 = 2l + 20 - 20$$
$$2l = 220$$
$$\frac{2l}{2} = \frac{220}{2}$$
$$l = 110 \text{ m}$$

Triangles

As we already have seen, any parallelogram is cut into two congruent triangles by either diagonal. This will prove very useful in the sections that follow. In the case of the rectangle that precedes, it is cut into two congruent right triangles.

Finding the Perimeter

For any triangle, $P_{\text{triangle}} = a + b + c$, where a, b, and c are the three sides of the triangle.

Consider $\triangle ABC$ with $a = 4$ in., $b = 7$ in., and $c = 10$ in. To find its perimeter:

$$P_{\text{triangle}} = a + b + c$$
$$P_{\text{triangle}} = 4 + 7 + 10$$
$$P_{\text{triangle}} = 11 + 10$$
$$P_{\text{triangle}} = 21 \text{ in.}$$

Finding the Area

Check out the preceding rectangle. When dealing with figures other than rectangles, we refer to one side as its base, and to a side perpendicular to the base as its height. Suppose that the rectangle has a base of 10 cm and a height of 7 cm. What is the area of either of the two triangles shown in that diagram?

First, let's find the area of the rectangle:

$$A_{\text{rectangle}} = lw \text{, which we'll alter to read } A_{\text{rectangle}} = bh$$
$$A_{\text{rectangle}} = (10)(7)$$
$$A_{\text{rectangle}} = 70 \text{ cm}^2$$

Well, it's clear that either triangle's area is half that of the rectangle, or 35 cm^2. That leads us to the formula:

$$A_{\text{triangle}} = \frac{1}{2}bh$$

Now it's clear that this formula works if the triangle in question is half a rectangle, but what if the triangle is one like $\triangle ABC$, as follows?

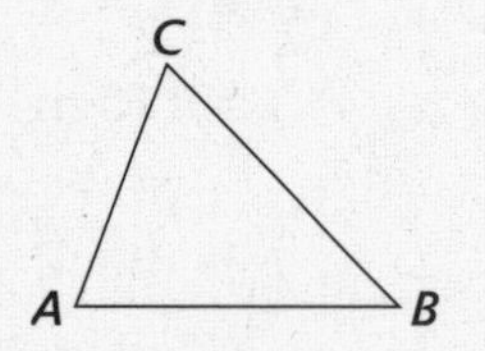

We'll examine that very good question right now.

Let's drop an altitude to base $\overline{AB}$ from C. That gives us $\overline{CD}$, the height of $\triangle ABC$, relative to base $\overline{AB}$, as you can see in Figure I, in the following diagram.

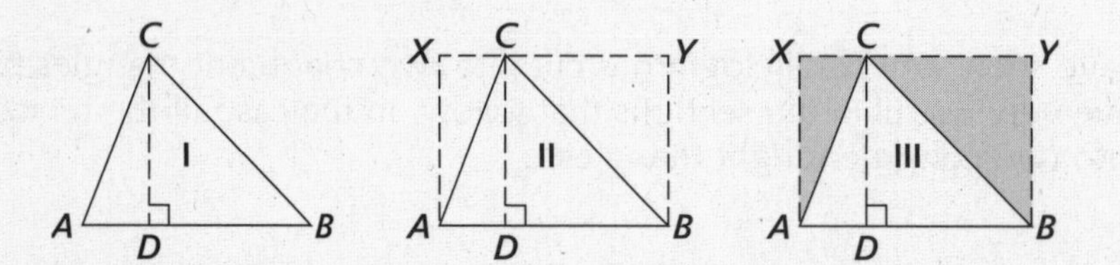

Next, we'll draw $\overline{AX}$, $\overline{XY}$, and $\overline{YB}$ to completely enclose $\triangle ABC$ in a rectangle (Figure II). Notice that $\overline{AC}$ is now dividing rectangle $ADCX$ into two congruent triangles, and $\overline{BC}$ divides $BDCY$ into two congruent triangles.

Now look at Figure III. The shaded sections added together form a triangle congruent to triangle ABC, because when equals are added to equals, the results are equal. That means that

$$A\triangle_{ABC} = \frac{1}{2}bh$$

I guess the triangle area formula works for any triangle!

Example Problems

These problems show the answers and solutions.

$\triangle DEF$ is a right triangle with legs 5 and 12 cm long and a hypotenuse of 13 cm. Use it to solve 1 and 2.

1. Find the area of $\triangle DEF$.

***Answer:* 30 cm²** The base and height of the triangle must be perpendicular to each other. In the case of a right triangle, that means the two legs. Remember: Area is always in square units.

$$A_{\text{triangle}} = \frac{1}{2}bh$$
$$A_{\text{triangle}} = \frac{1}{2}(12)(5)$$
$$A_{\text{triangle}} = \frac{1}{2}(60)$$
$$A_{\text{triangle}} = 30\,\text{cm}^2$$

2. Find the perimeter of $\triangle DEF$.

***Answer:* 30 cm**

$P_{\text{triangle}} = d + e + f$ (We don't know which side is which, but it doesn't matter.)
$P_{\text{triangle}} = 5 + 12 + 13$
$P_{\text{triangle}} = 17 + 13$
$P_{\text{triangle}} = 30$ cm

3. In $\triangle MNO$, $NO = 8$ m, $MO = 9$ m, $MN = 12$ m. The altitude to O is 6. Find the area of $\triangle MNO$.

***Answer:* 36 m²** If you draw a picture, you'll see that the only height you have is the altitude to O, which is 6 m long. That makes side O the default base. The other two measurements are irrelevant.

$$A_{\text{triangle}} = \frac{1}{2}bh$$
$$A_{\text{triangle}} = \frac{1}{2}(12)(6)$$
$$A_{\text{triangle}} = \frac{1}{2}(72)$$
$$A_{\text{triangle}} = 36\,\text{m}^2$$

Parallelograms

Parallelograms have characteristics in common with the rectangle (after all, a rectangle is a parallelogram) and characteristics in common with triangles—nonright angles. You'll see that these characteristics affect the special area and perimeter formulas for parallelograms.

Finding the Perimeter

In common with a rectangle, a parallelogram has two pairs of nonadjacent equal sides.

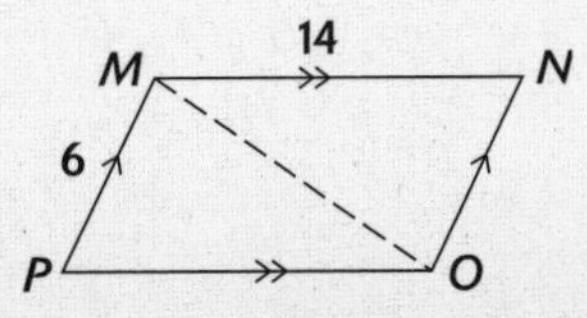

Call the horizontal sides *a* and the oblique sides *b*, and it is obvious that

$$P_{\text{parallelogram}} = 2(a + b)$$

To find the perimeter of ▱*MNOP*,

$$P_{\text{parallelogram}} = 2(a + b)$$
$$P_{\text{parallelogram}} = 2(14 + 6)$$
$$P_{\text{parallelogram}} = 2(20)$$
$$P_{\text{parallelogram}} = 40$$

Finding the Area

Notice that in the ▱*MNOP*, the diagonal divides the figure into two congruent triangles. We learned that before, but it bears repeating here. Since we know that the area of a triangle is $\frac{1}{2}bh$, it stands to reason that a parallelogram's area is twice that, or *bh*. Here's another way to demonstrate that.

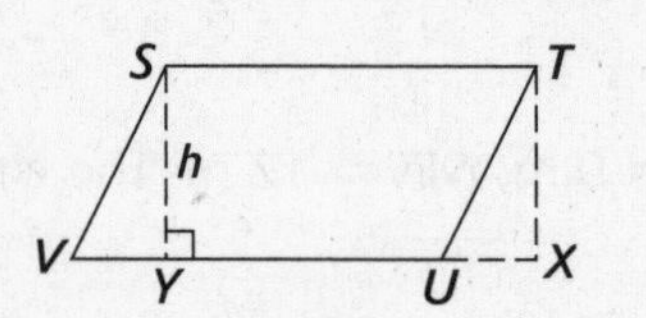

In ▱*STUV*, the height, $\overline{SY}$, has been drawn. Together with $\overline{VY}$ and $\overline{SV}$, it forms △*SVY*. If we cut that triangle out of the figure and move it to the right, and rename it △*TUX*, we form rectangle *STXY*. Clearly, the area of that rectangle is its base, $\overline{XY}$ by its height, $\overline{SY}$, so for either of the two reasons given previously, $A_{\text{parallelogram}} = bh$.

Example Problems

These problems show the answers and solutions.

▱*ABCD* has an obtuse angle at *C*. *CD* = 18 cm, *BC* = 10 cm. Height *CX* is 6 cm and is perpendicular to $\overline{AB}$. Use this figure to solve problems 1 and 2.

1. Find the area of ▱*ABCD*.

   ***Answer:* 108 cm²** Since *C* is an obtuse angle and *CX* is the height. You really should draw the diagram before solving.

$$A_{\text{parallelogram}} = bh$$
$$A_{\text{parallelogram}} = 18 \times 6$$
$$A_{\text{parallelogram}} = 108\text{ cm}^2$$

2. Find the perimeter of ▱*ABCD*.

***Answer:* 56 cm**

$$P_{\text{parallelogram}} = 2(a + b)$$
$$P_{\text{parallelogram}} = 2(18 + 10)$$
$$P_{\text{parallelogram}} = 2(28)$$
$$P_{\text{parallelogram}} = 56 \text{ cm}$$

3. The area of ▱*PQRS* is 48 ft^2 Its height is a whole number. *QR* is 8 feet; *SR* is 11 ft. What is the height?

***Answer:* 6 ft.**

$$A_{\text{parallelogram}} = bh$$

A base of 11 ft. will not divide perfectly into 48 ft^2

$$48 = 8h$$
$$8\frac{h}{8} = \frac{48}{8}$$
$$h = 6 \text{ ft.}$$

Work Problems

Use these problems to give yourself additional practice. Draw your own diagrams.

1. A parallelogram has bases of 13 feet and 21 feet. What is its perimeter?

2. $\triangle ABC$ has obtuse $\angle B$. $AB = 17$ in.; $BC = 24$ in. Height AY is 9 in. and meets BC at Y. What is the area of $\triangle ABC$?

3. The perimeter of $\triangle RST$ is 118 cm. Side $ST = 42$ cm, and $RS = 39$ cm. Find side *S*.

4. The area of ▱*WXYZ* is 342 m^2, and its height is 9. Find its base.

5. A right triangular parking lot has an area of 645 square meters. The shortest side of the parking lot is 15 meters. What is the length of the second shortest side?

Worked Solutions

1. **68 ft.**

$$P_{\text{parallelogram}} = 2(a + b)$$
$$P_{\text{parallelogram}} = 2(13 + 21)$$
$$P_{\text{parallelogram}} = 2(34)$$
$$P_{\text{parallelogram}} = 68 \text{ ft.}$$

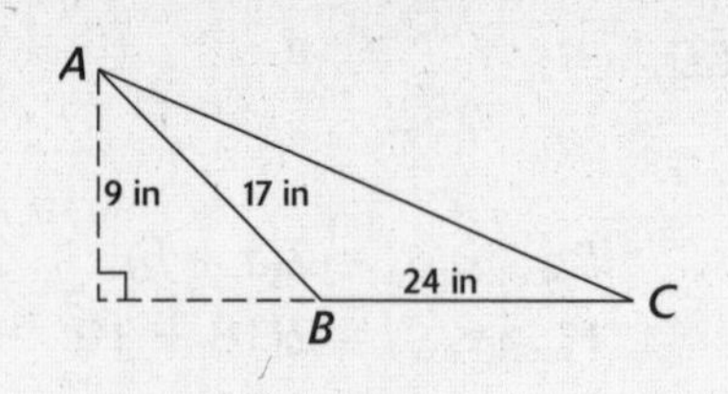

2. **108 in²**

$$A_{\text{triangle}} = \frac{1}{2}bh$$
$$A_{\text{triangle}} = \frac{1}{2}(24)(9)$$
$$A_{\text{triangle}} = \frac{1}{2}(216)$$
$$A_{\text{triangle}} = 108\text{in}^2$$

3. **37 cm**

$$P_{\text{triangle}} = ST + RT + RS$$
$$118 = 42 + S + 39$$
$$118 = 81 + S$$
$$118 - 81 = 81 - 81 + S$$
$$S = 37 \text{ cm}$$

4. **38 m**

$$A_{\text{parallelogram}} = bh$$
$$342 = 9b$$
$$\frac{9b}{9} = \frac{342}{9}$$
$$b = 38 \text{ m}$$

5. **86 m** Remember, the two shorter sides of a right triangle are its legs. Which are ⊥?

$$A_{\text{triangle}} = \frac{1}{2}bh$$
$$645 = \frac{1}{2}15b$$
$$2(645) = 2\left(\frac{1}{2}\right)15b$$
$$\frac{15b}{15} = \frac{1290}{15}$$
$$b = 86\text{m}$$

Trapezoids

The trapezoid is probably the most popular quadrilateral in the field of engineering. You can recognize its shape in traffic bridges and railroad trestles built from the late nineteenth through the mid-twentieth centuries. Keep an eye out for one, because they're practically everywhere.

Finding the Perimeter

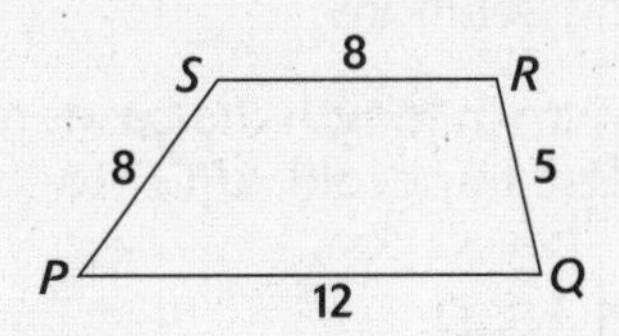

Finding the perimeter of a trapezoid is just as easy as a + b + c + d. Find the perimeter of trapezoid *PQRS* (in the preceding figure). The dimensions shown are in inches.

$P_{\text{trapezoid}} = a + b + c + d$ (No particular order is required.)

$P_{\text{trapezoid}} = 12 + 5 + 8 + 8$

$P_{\text{trapezoid}} = 17 + 16$

$P_{\text{trapezoid}} = 33$ in.

Finding the Area

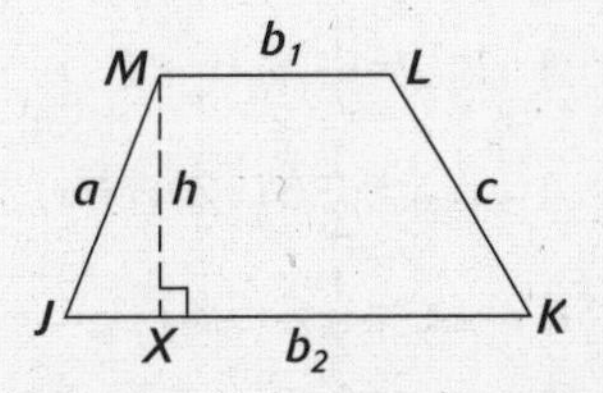

There is a special formula for finding the areas of trapezoids. Take a look at the preceding figure and notice that we have labeled the lengths of the two parallel bases as b_1 and b_2. b_2 is the entire length of $\overline{JK}$. The legs are of lengths *a* and *c*, respectively, and the height is *h*. There is more than one way to derive the formula for the area of a trapezoid, but we're going to go with the "make a parallelogram" idea. Starting out with trapezoid *JKLM*, as shown, we draw the identical trapezoid, inverted, and join it to *JKLM* to form $\square$*JPNM*, as shown here:

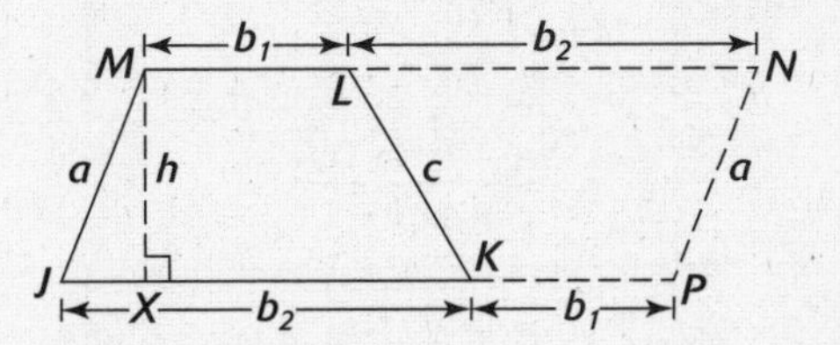

You should realize that since $\square$*JPNM* contains two identical trapezoids, its area must be twice that of one of them. Using the area of a parallelogram formula, let's find the area of $\square$*LPNM*.

$A_{\text{parallelogram}} = bh$

$A_{\text{parallelogram}} = (b_1 + b_2)h$ or $A_{\text{parallelogram}} = h(b_1 + b_2)$

Therefore, $A_{\text{trapezoid}} = \frac{1}{2}h(b_1 + b_2)$.

Example Problems

These problems show the answers and solutions.

Trapezoid *ABCD* has legs 7 cm and 9 cm in length. Its bases are 12 cm and 16 cm, respectively. The height of the trapezoid is 6 cm. Use trapezoid *ABCD* to solve problems 1 and 2.

1. Find the perimeter of trapezoid *ABCD*.

Answer: 44 cm

$$P_{\text{trapezoid}} = a + b + c + d$$
$$P_{\text{trapezoid}} = 7 + 9 + 12 + 16$$
$$P_{\text{trapezoid}} = 16 + 28$$
$$P_{\text{trapezoid}} = 44 \text{ cm}$$

2. Find the area of trapezoid *ABCD*.

Answer: 84 cm²

$$A_{\text{trapezoid}} = \frac{1}{2}h(b_1 + b_2)$$
$$A_{\text{trapezoid}} = \frac{1}{2}6(12 + 16)$$
$$A_{\text{trapezoid}} = \frac{1}{2}6(28)$$
$$A_{\text{trapezoid}} = 3(28)$$
$$A_{\text{trapezoid}} = 84 \text{ cm}^2$$

3. The strange-looking figure that follows is a trapezoidal-shaped play mat. Find its area.

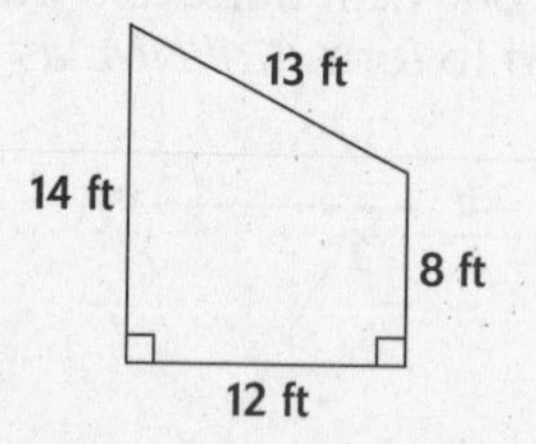

Answer: 132 ft²

$$A_{\text{trapezoid}} = \frac{1}{2}h(b_1 + b_2)$$
$$A_{\text{trapezoid}} = \frac{1}{2}12(8 + 4)$$
$$A_{\text{trapezoid}} = 6(22)$$
$$A_{\text{trapezoid}} = 132 \text{ ft}^2$$

Regular Polygons

So far, we have studied figures with three and four sides. The fact of the matter is that there is no limit to the number of sides a regular figure may have, although on an ordinary sheet of paper, after you've gotten up to about 20 to 25 sides, everything starts to look pretty much like a circle.

Special Parts of Regular Polygons

Each regular polygon has one point that is an equal distance from all of its vertices. That point is known as the **center** of the polygon, and is shown at O in the hexagon that follows.

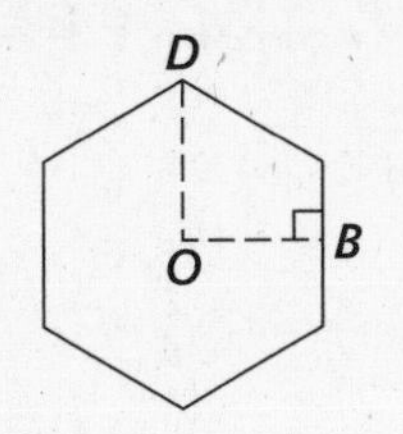

$\overline{OD}$ is a **radius** of this polygon. It connects the center to a vertex. A polygon has as many radii as it has vertices.

$\overline{OB}$ is an **apothem**. It is drawn from the center and is perpendicular to one of the polygon's sides. A polygon will have as many apothems as it has sides.

Finding the Perimeter

Since a regular polygon is equilateral, to find its perimeter, all that is needed is to know the length of one of its sides, s, and the number of sides, n, that the polygon has. If we use the expression n-gon to represent a polygon of n sides, then the formula for the perimeter is as follows:

$$P_{\text{regular } n\text{-gon}} = ns$$

Finding the Area

If we use p to represent the perimeter of a regular polygon and a to represent the length of its apothem, then it is possible to prove (by way of a whole lot of internal congruent triangles the polygon can be broken up into) that the following formula represents its area:

$$A_{\text{regular } n\text{-gon}} = \frac{1}{2}ap$$

Example Problems

These problems show the answers and solutions.

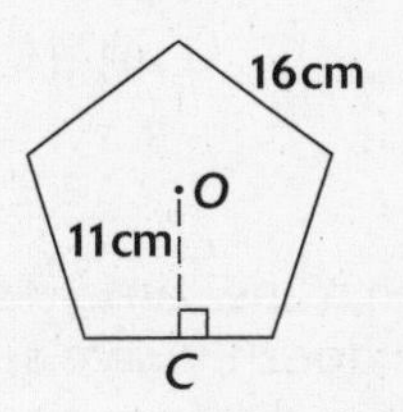

1. The regular pentagon shown has sides 16 cm long and an apothem of *about** 11 cm. Find its area.

   ***Answer:* 440 cm²**

To begin, we'll need to find the perimeter:

$$P_{\text{regular } n\text{-gon}} = ns$$
$$P_{\text{regular } n\text{-gon}} = 5 \cdot 16$$
$$P_{\text{regular } n\text{-gon}} = 80 \text{ cm}$$

Now, moving right along:

$$A_{\text{regular } n-\text{gon}} = \frac{1}{2}ap$$
$$A_{\text{regular } n-\text{gon}} = \frac{1}{2}(11 \cdot 80)$$
$$A_{\text{regular } n-\text{gon}} = \frac{1}{2}(880)$$
$$A_{\text{regular } n-\text{gon}} = 440 \text{ cm}^2$$

Now how much fun was that?!

(*If the actual apothems and sides were included, this and other n-gon problems would be complicated by decimal fractions. Since the purpose is to learn the concept and not to practice complicated decimal arithmetic, we've rounded to whole numbers.)

2. A regular octagonal rug has sides 4 ft long and an apothem of about 3 ft. Find its area.

***Answer:* 48 ft²**

To begin, we find the perimeter:

$$P_{\text{regular } n\text{-gon}} = ns$$
$$P_{\text{regular } n\text{-gon}} = 8 \cdot 4$$
$$P_{\text{regular } n\text{-gon}} = 32 \text{ ft.}$$

Now, moving right along:

$$A_{\text{regular } n-\text{gon}} = \frac{1}{2}ap$$
$$A_{\text{regular } n-\text{gon}} = \frac{1}{2}(3 \cdot 32)$$
$$A_{\text{regular } n-\text{gon}} = \frac{1}{2}(96)$$
$$A_{\text{regular } n-\text{gon}} = 48 \text{ ft}^2$$

Circles

A circle is a plane figure with all points on it the same distance from a single fixed point, known as the **center of the circle**. Since a circle doesn't have segments for sides, we need a special name for its fence, or perimeter. We call the distance around the circle its **circumference**.

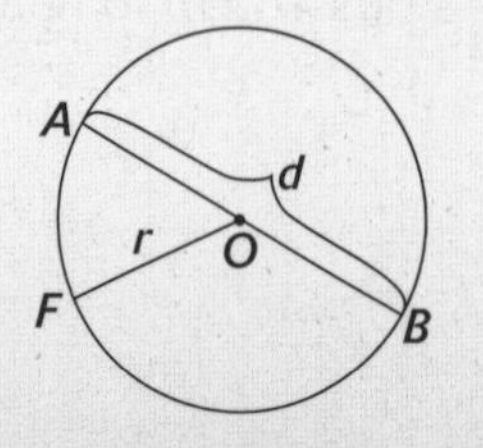

Any segment that connects the center of the circle to the circumference is called a **radius**. $\overline{OF}$ is a radius of circle O. There is no limit to the number of **radii** (plural of radius) that a circle may have. A line segment that begins on the circumference, passes through the center, and ends on the opposite side of the circumference is a **diameter**. $\overline{AB}$ is a diameter of circle O. Obviously, in the same or equal circles, a diameter is the same length as two radii. No line segment inside a circle is larger than the diameter, and no limit exists to the number of diameters that a circle may have.

Finding Circumference

More than two thousand years ago, the Greeks discovered that the circumference of a circle is 3 and a little bit more times the length of a diameter. The Greeks represented this amount with the letter π, (called pi and pronounced just like pie). π is a constant, usually represented by the decimal 3.14, or the fraction $\frac{22}{7}$. The circumference of a circle may be represented by either of the following:

$$C = \pi d \quad \text{or} \quad C = 2\pi r$$

Since d and $2r$ mean exactly the same thing, the length of a diameter, there's no need to memorize both formulas. We suggest learning the first, since the second might get confused with the formula for the area of a circle, which we'll get into in a bit.

Example Problems

These problems show the answers and solutions. Leave answers in terms of pi, where appropriate.

1. A circle has a diameter of 2 m. What is its circumference?

***Answer:* 2π m**

$$C = \pi d$$
$$C = \pi \cdot 2$$
$$C = 2\pi \text{ m}$$

2. A circle has a radius of 5 inches. What is its circumference?

***Answer:* 10π in.**

$$C = \pi d$$
$$C = \pi \cdot 10 \quad \text{Remember: } d = 2r.$$
$$C = 10\pi \text{ in.}$$

3. A traffic circle's circumference is 300π yards. What is its radius?

***Answer:* 150 yd.**

$$C = \pi d$$
$$300\pi = \pi d$$
$$300\pi = \pi d$$
$$\frac{300\pi}{\pi} = \frac{\pi d}{\pi}$$

$$d = 300 \quad \text{(But the question asked for the radius.)}$$
$$r = \frac{1}{2}d$$
$$r = \frac{1}{2} \cdot 300$$
$$r = 150 \text{ yd.}$$

Finding the Area

The formula for the area of a circle can be derived from the area of a polygon, if the apothem and radius of a circle are considered to be the same length, but why bother? If you're interested, look at the derivation below.* If you don't care, skip to the Example Problems.

*Here is how A_{circle} is derived from $A_{\text{regular } n\text{-gon}}$. The circle's apothem coincides with the radius.

$$A_{\text{regular } n-gon} = \frac{1}{2}ap$$
$$A_{\text{circle}} = \frac{1}{2}(r)(2\pi r)$$
$$A_{\text{circle}} = \frac{1}{2}(2\pi r)(r)$$
$$A_{\text{circle}} = \pi r^2$$

To find the area of a circle, the radius is multiplied by itself, and the result multiplied by π. The formula looks like this:

$$A_{\text{circle}} = \pi r^2$$

The result, as with all area problems, is expressed in square units.

Example Problems

These problems show the answers and solutions.

1. Find the area of a circle of radius 7 cm.

***Answer:* 49π cm²**

$$A_{\text{circle}} = \pi r^2$$
$$A_{\text{circle}} = \pi \cdot 7^2$$
$$A_{\text{circle}} = 49\pi \text{ cm}^2$$

2. Find the area of a circle of diameter 20 yd.

***Answer:* 100π yd²**

$$A_{\text{circle}} = \pi r^2$$
$$A_{\text{circle}} = \pi \cdot 10^2 \quad \text{We were given } d. \quad r = \frac{d}{2}$$
$$A_{\text{circle}} = 100\pi \text{ yd}^2$$

3. A circular swimming pool has a surface area of 400π ft² What is the diameter of the pool?

Answer: **40 ft.**

$A_{circle} = \pi r^2$	
$400\pi = \pi r^2$	We already know the area, so substitute it.
$\frac{400\pi}{\pi} = \frac{\pi r^2}{\pi}$	We divide both sides by π.
$400 = r^2$	Next we take the square root of both sides.
$\sqrt{400} = \sqrt{r^2}$	
$r = 20$	It's turned around, but it means the same as $20 = r$.
$d = 40$ ft.	The question asked for the diameter ($2r$).

Work Problems

Use these problems to give yourself additional practice.

1. The widest distance across a carousel is 35 feet. What is the distance around the carousel?

2. A car's tire has a radius of 7.5 inches. How far will the car have moved after the tire has turned exactly one revolution?

3. A circle has a radius of 17 cm. What is the circle's area?

4. A circus ring is 48 feet across. How much of a surface is there for the performers' use?

5. A dome is in the shape of a semi-circle (that's a half circle). Its cross section at the widest part has an area of 72π m². What is the longest distance across the bottom of the dome?

Worked Solutions

1. **35π ft.** A carousel is round. The "widest distance across" it is its diameter. The "distance around" is the circumference.

$C = \pi d$	Use the circumference formula.
$C = \pi \cdot 35$	Substitute 35 for d.
$C = 35\pi$ ft.	Solve.

2. **$C = 15\pi$ in.**

$C = \pi d$	After one revolution, the car will have moved the length of the tire's circumference.
$C = \pi \cdot 15$	Remember: $d = 2r$; $2 \cdot 7.5 = 15$
$C = 15\pi$ in.	

3. **289π cm^2**

$A_{circle} = \pi r^2$ This is a straightforward area of a circle problem.
$A_{circle} = \pi \cdot 17^2$ We plug in the radius. . .
$A_{circle} = 289\pi$ cm^2 . . . and solve.

4. **576π ft^2**

$A_{circle} = \pi r^2$ A ring is a circle; the inside surface is its area.
$A_{circle} = \pi \cdot 24^2$ We were given d. $r = \frac{d}{2}; \frac{48}{2} = 24$
$A_{circle} = 576\pi$ ft^2

5. **24 m** Since the cross section is that of a half circle, the area of the full circle would be twice that, or 144π m^2. Substitute that into the area formula, and we'll go from there.

$A_{circle} = \pi r^2$
$144\pi = \pi r^2$ We substitute the area.
$\frac{144\pi}{\pi} = \frac{\pi r^2}{\pi}$ We divide both sides by π.
$144 = r^2$ Next we'll take the square root of both sides.
$\sqrt{144} = \sqrt{r^2}$
$r = 12$ m It's turned around, but it means the same as $12 = r$.

The question asked for the distance across the widest part. That's the circle's diameter:

$d = 24$ m The diameter is twice the radius ($2r$).

Chapter 6

Similar Figures

Earlier, we studied congruent figures—those with the same shape and size. Similar figures are not usually the same size, but they are identical in shape. The secret to having an identical shape is having all angles identical in degree measure. Studying triangles with identical angles but different side-lengths led to the discovery of the trigonometric functions and gave birth to a whole new branch of mathematics, but that's the subject for a different book.

Ratio and Proportion

No doubt you have run across ratio before in a math class. *Ratio* is a mathematical name for a comparison. Many quantities may be compared to each other, but for all practical purposes in geometry, ratios are used to compare two quantities.

Ratio

Essentially, you have three ways to represent the ratio of quantity a to quantity b. We may write a to b; $a{:}b$, or $\frac{a}{b}$. When three or more quantities are compared, the colon form is preferred ($a{:}b{:}c$). When writing ratios in fractional form, it is customary to express the fraction in lowest terms.

For any pair of quantities, it is possible to write six ratios. Consider a bowl containing 6 pears and 5 apples:

The ratio of pears to apples is $\frac{6}{5}$.

The ratio of apples to pears is $\frac{5}{6}$.

The ratio of pears to pieces of fruit is $\frac{6}{11}$.

The ratio of apples to pieces of fruit is $\frac{5}{11}$.

The ratio of pieces of fruit to pears is $\frac{11}{6}$.

The ratio of pieces of fruit to apples is $\frac{11}{5}$.

Now that's what we call raw data. It has a certain uh, peel!

Example Problems

These problems show the answers and solutions.

1. A bus contains 18 boys and 24 girls. What is the ratio of boys to girls?

   ***Answer:* 3:4 or $\frac{3}{4}$** The ratio of boys to girls is 18 to 24, 18:24, or $\frac{18}{24}$. Since the GCF (you remember Greatest Common Factor, don't you?) is 6, $\frac{18}{24}$ simplifies to $\frac{3}{4}$.

2. The 100 foot length of rope is divided into three parts in the ratio 2:3:5. How long is the middle-sized length of rope?

   ***Answer:* 30 ft.** Let the measure of the shortest piece be $2l$. That makes the middle-sized piece $3l$, and the longest piece $5l$. Then:

   $$2l + 3l + 5l = 100$$
   $$10l = 100$$
   $$l = 10$$

   That makes the pieces 20, 30, and 50 feet. The middle-sized piece is 30 ft.

3. Two supplementary angles are in the ratio 5:7. How large is each angle?

   ***Answer:* 75° and 105°**

   Let $5d$ = the smaller angle and $7d$ = the larger angle. Supplementary angles, you'll recall, total 180°.

   $$5d + 7d = 180°$$
   $$12d = 180°$$
   $$d = 15°$$
   $$5d = 5 \cdot 15 \qquad 7d = 7 \cdot 15$$
   $$5d = 75° \qquad 7d = 105°$$

Proportions

A proportion is an equation that states that two different ratios are equal. Here are two proportions:

$$10{:}25 = 2{:}5 \qquad \frac{8}{24} = \frac{1}{3}$$

Means and Extremes

The **means** of a proportion are those terms that are closest together. In the case of $a{:}b = c{:}d$, b and c are the means. The terms farthest apart, a and d are known as the **extremes**.

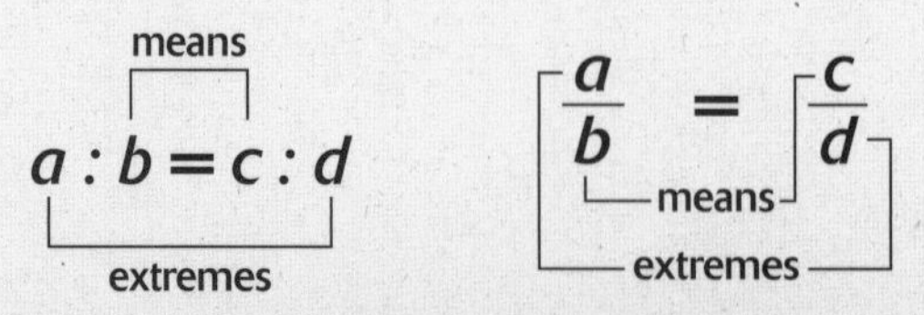

Study the diagram shown here. It shows the means and extremes of a proportion, bracketed both in the colon format and in the fraction format.

Properties of Proportions

The following four properties of proportions can be proved easily algebraically, but we will only state them here. The first one is by far the most important.

1. **The Cross-Products Property (or The Means-Extremes Property):**

   If $\frac{a}{b} = \frac{c}{d}$, then $ad = bc$.
   Conversely, if $ad = bc \neq 0$, then $\frac{a}{b} = \frac{c}{d}$, and $\frac{b}{a} = \frac{d}{c}$.

   $$\frac{a}{b} \times \frac{c}{d} \longrightarrow ad = bc$$

   $\frac{2}{4} = \frac{6}{12}$ is a proportion.

   By the Cross-Products Property, $(2)(12) = (4)(6)$.
   That means: $24 = 24$.

   Using the cross-product property, we not only can find a missing term in a proportion. We also can find out whether an equation is a proportion. Consider the following:

   $\frac{3}{4} = \frac{5}{6}$ Is this a true proportion?

   To test, cross multiply: $(3)(6) = (4)(5)$
   $18 = 20$

   Does 18 = 20? Of course not. Therefore, $\frac{3}{4} = \frac{5}{6}$ is *not* a true proportion.

2. **The Means or Extremes Switching Property:**

   If $\frac{a}{b} = \frac{c}{d}$ and is a true proportion, then both $\frac{d}{b} = \frac{c}{a}$, and $\frac{a}{c} = \frac{b}{d}$ are true proportions.

   This means that in the proportion $\frac{3}{6} = \frac{1}{2}$, the extremes 3 and 2 may be switched to make $\frac{2}{6} = \frac{1}{3}$, or the means 1 and 6 may be switched to make $\frac{3}{1} = \frac{6}{2}$, and both those would still be true proportions. Check them out, and you'll see that they are, in fact, true proportions.

3. **The Upside Down Property:**

   If you turn a true proportion over, it is also a true proportion. That is: If $\frac{a}{b} = \frac{c}{d}$, then $\frac{b}{a} = \frac{d}{c}$.

   Check this out:

   $\frac{a}{5} = \frac{8}{10}$ upside down is $\frac{5}{a} = \frac{10}{8}$
   $10a = 40$ $10a = 40$
   $a = 4$ $a = 4$

4. The Denominator Addition or Subtraction Property:

If $\frac{a}{b}=\frac{c}{d}$, then $\frac{(a+b)}{b}=\frac{(c+d)}{d}$ or $\frac{(a-b)}{b}=\frac{(c-d)}{d}$.

To test addition:

$$\frac{8}{12}=\frac{2}{3} \qquad \frac{(8+12)}{12}=\frac{(2+3)}{3}$$

$$8\cdot 3=2\cdot 12 \qquad 20/12=5/3$$

$$24=24 \qquad 20\cdot 3=12\cdot 5$$

$$60=60$$

To test subtraction:

$$\frac{8}{12}=\frac{2}{3} \qquad \frac{(8-12)}{12}=\frac{(2-3)}{3}$$

$$8\cdot 3=2\cdot 12 \qquad \frac{-4}{12}=\frac{-1}{3}$$

$$24=24 \qquad (-4)(3)=(12)(-1)$$

$$-12=-12$$

Well, it looks like both the addition and subtraction parts of the fourth property work.

Example Problems

These problems show the answers and solutions.

1. Use the cross-product property to find the value of c in the proportion $\frac{c}{7}=\frac{28}{49}$.

***Answer:* 4**

$$49c=7\cdot 28$$

$$49c=196$$

$$\frac{49c}{49}=\frac{196}{49}$$

$$c=4$$

2. $\frac{a}{5}=\frac{b}{9}$ is a proportion. Find the value of $\frac{a}{b}$.

Answer: $\mathbf{\frac{5}{9}}$ To solve, we use the means or extremes switching property (the second property) to switch the means, 5 and b. That changes the proportion to $\frac{a}{b}=\frac{5}{9}$—the solution.

3. If $\frac{7}{x}=\frac{9}{y}$, find the value of $\frac{x}{y}$.

Answer: $\mathbf{\frac{7}{9}}$ First, use the means or extremes switching property (the second property) to switch the extremes, 7 and y. That changes the proportion to $\frac{y}{x}=\frac{9}{7}$. Next, use the upside-down property to flip the proportion over and get $\frac{x}{y}=\frac{7}{9}$.

4. In $\triangle ABC$, in the following figure, $\frac{BE}{CE} = \frac{3}{4}$. Find $\frac{BC}{CE}$.

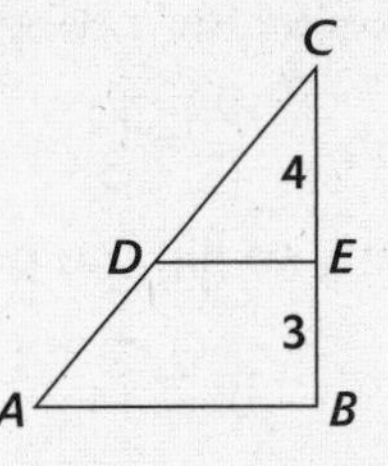

Answer: $\frac{7}{4}$ By the Denominator Addition (or Subtraction) Property:

If $\frac{a}{b} = \frac{c}{d}$ then $\frac{(a+b)}{b} = \frac{(c+d)}{d}$.

Therefore: Since $\frac{BE}{CE} = \frac{3}{4}$, $\frac{(BE+CE)}{CE} = \frac{(3+4)}{4}$, but $(BE + CE) = BC$, so:

$$\frac{BC}{CE} = \frac{7}{4}$$

Work Problems

Use these problems to give yourself additional practice.

1. Find the value of x in the proportion $\frac{x}{9} = \frac{63}{81}$.

2. If $\frac{5}{x} = \frac{11}{y}$, find the value of $\frac{x}{y}$.

3. $\frac{c}{3} = \frac{d}{8}$ is a proportion. Find the value of $\frac{c}{d}$.

4. In $\square ABCD$, E is on side CD, 4 cm from C and 5 cm from D. $\frac{CE}{DE} = \frac{4}{5}$. Find $\frac{CD}{DE}$.

5. The sides of a rectangle are in the ratio 3:5. If a short side of the rectangle is 24 inches long, what is the length of a long side?

Worked Solutions

1. **7** We use the cross-product property to find the value of x in the proportion.

$$\frac{x}{9} = \frac{63}{81}$$
$$81x = 9 \cdot 63$$

Divide both sides by 9, since $81 = 9 \cdot 9$.

$$9x = 63$$
$$\frac{9x}{9} = \frac{63}{9}$$
$$x = 7$$

2. $\frac{5}{11}$

First, use the means or extremes switching property to switch the extremes, 5 and y.

$$\frac{y}{x} = \frac{11}{5}$$

Next, use the upside-down property to flip the proportion over:

$$\frac{x}{y} = \frac{5}{11}$$

3. $\frac{3}{8}$ Solve using the means or extremes switching property to switch the means, 3 and d. That changes the proportion to $\frac{c}{d} = \frac{3}{8}$, which is the solution.

4. $\frac{9}{5}$ Using the denominator addition property: If $\frac{a}{b} = \frac{c}{d}$, then $\frac{(a+b)}{b} = \frac{(c+d)}{d}$.

Therefore: Since $\frac{CE}{DE} = \frac{4}{5}$,

$$\frac{(CE + DE)}{DE} = \frac{(4+5)}{5} \qquad \frac{CD}{DE} = \frac{9}{5}$$

5. **40 in.**

First make a proportion: $\frac{s}{l} = \frac{3}{5}$

Next substitute: $\frac{24}{l} = \frac{3}{5}$

Cross multiply: $3l = 5 \cdot 24 = 120$

Divide both sides by 3: $\frac{3l}{3} = \frac{120}{3}$

$l = 40$

Don't forget to express the answer in inches: 40 in.

Similar Polygons

Polygons with the same shape are known as **similar polygons**. By the same shape, we don't mean all rectangles or all triangles. To be similar, all pairs of corresponding sides must be in proportion, **and** each pair of corresponding angles must be equal. It is not enough for just one of those conditions to be true. For example, look at the two quadrilaterals that follow.

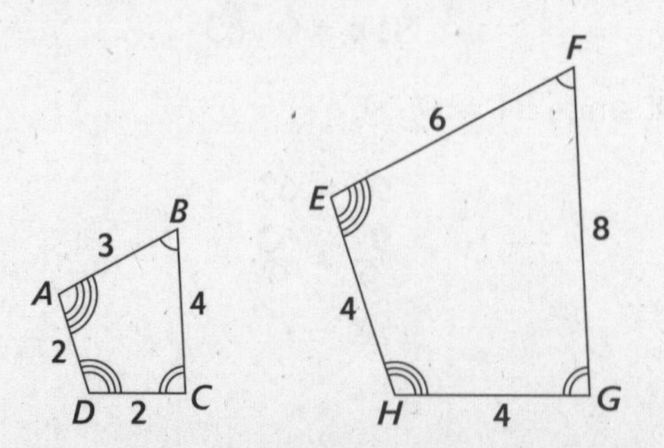

Quadrilateral $ABCD$ ~ quadrilateral $EFGH$. (Read that "~" as "is similar to.") That means $m\angle A = m\angle E$, $m\angle B = m\angle F$, $m\angle C = m\angle G$, $m\angle D = m\angle H$, **and** $\frac{AB}{EF} = \frac{BC}{FG} = \frac{CD}{GH} = \frac{AD}{EH}$.

Example Problems

These problems show the answers and solutions. Tell whether the figures in each pair are similar. If so, tell why; if not, tell why not.

1.

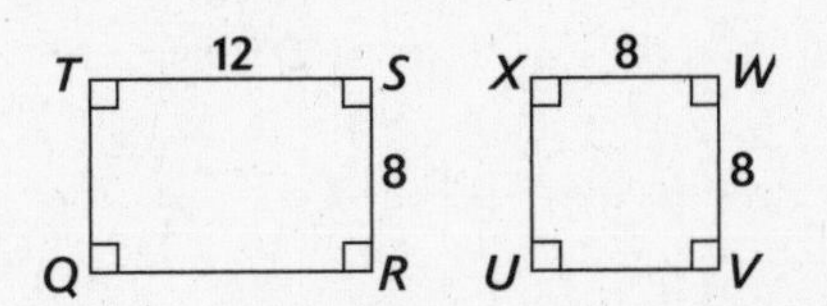

***Answer:* No** The corresponding angles are equal, but only two corresponding sides are in proportion. The vertical sides of both figures are in 1:1 ratio, but the horizontal sides are in 3:2 ratio. All must be in the same ratio for the figures to be similar.

2.

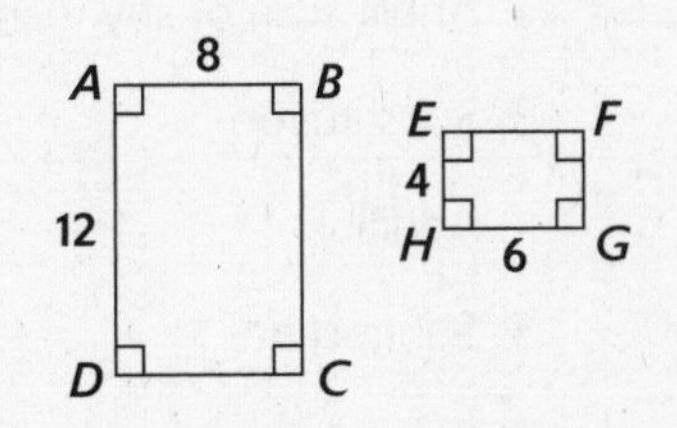

***Answer:* Yes** The corresponding angles are equal; after all, they're all right angles. The corresponding sides are all in 2:1 ratio: $\frac{AB}{EH} = \frac{BC}{EF} = \frac{CD}{FG} = \frac{AD}{GH} = \frac{1}{2}$.

3.

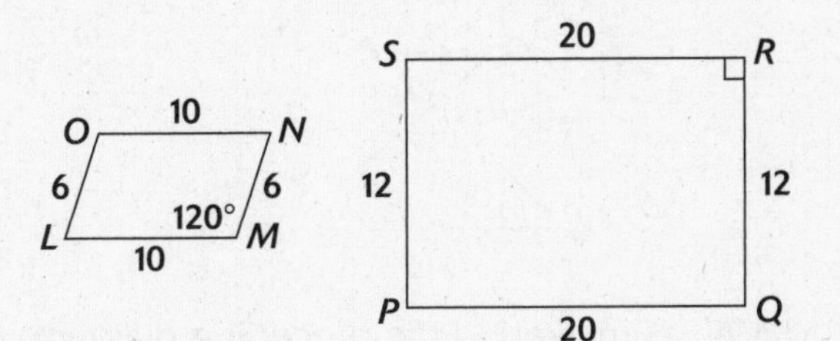

***Answer:* No** Here the corresponding sides are all in the ratio 1:2. That's a good sign. However, the first figure, ▱*LMNO* is a parallelogram with opposite angles of 120° and 60°. ▱*PQRS* is a rectangle with 90° angles. The corresponding angles are not equal, so the figures are not similar.

Similar Triangles

As a general rule, in order to prove that two polygons are similar, it is necessary to prove that all corresponding angles are equal and all corresponding sides are in proportion (that is the ratios between them are all equal). Triangles, however, are a special case.

Postulate 18: If two angles of one triangle are equal to two angles of a second triangle, then the triangles are similar. (This is known as the AA Triangle Similarity Postulate.)

Example Problems

These problems show the answers and solutions.

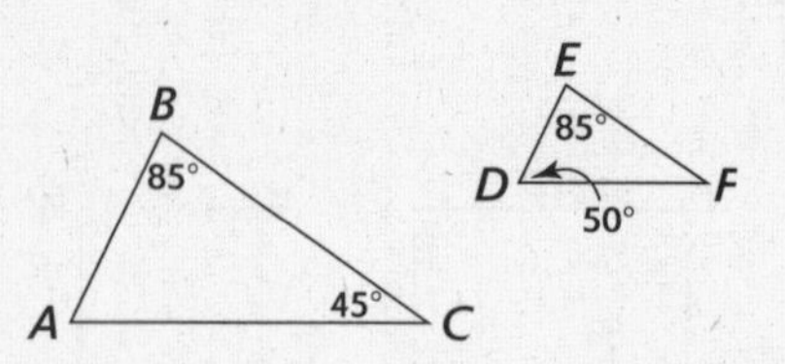

1. *Given:* $\triangle ABC$ and $\triangle DEF$, as marked in the preceding diagram. *Prove:* $\triangle ABC \sim \triangle FED$.

Statement	***Reason***
1. $m\angle B = m\angle E = 85°$	1. Given (as marked).
2. $m\angle A + m\angle B + m\angle C = 180°$	2. Angle sum of any triangle = 180°.
3. $m\angle A + 85° + 45° = 180°$	3. Substitution
4. $m\angle A + 130° = 180°$	4. Addition
5. $m\angle A = 50°$	5. Subtraction
6. $m\angle D = 50°$	6. Given (as marked).
7. $m\angle A = m\angle D$	7. Two angles equal to the same quantity are equal to each other.
8. $\triangle ABC \sim \triangle FED$	8. AA Similarity Postulate (18)

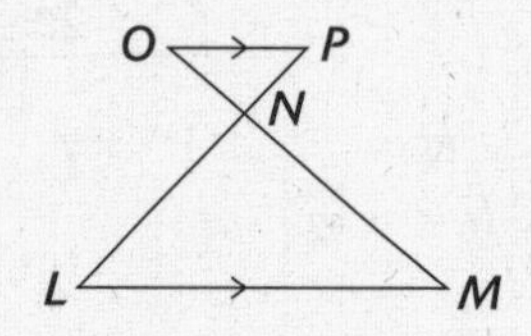

2. *Given:* $\triangle NOP$ and $\triangle LMN$, as marked in the preceding diagram. *Prove:* $\triangle NOP \sim \triangle NML$.

Statement	***Reason***
1. $\overline{OP} \parallel \overline{LM}$	1. Given (as marked).
2. $\angle M$, $\angle O$, $\angle L$, and $\angle P$ are pairs of alt. int. $\angle$s.	2. Definition of alternate interior angles of //
3. $m\angle M = m\angle O$ and $m\angle L = m\angle P$	3. Alternate interior angles are equal.
4. $\triangle NOP \sim \triangle NML$	4. AA Similarity Postulate (18)

One of the pairs of angles in step 3 could have been replaced by equating the two vertical angles at *N*.

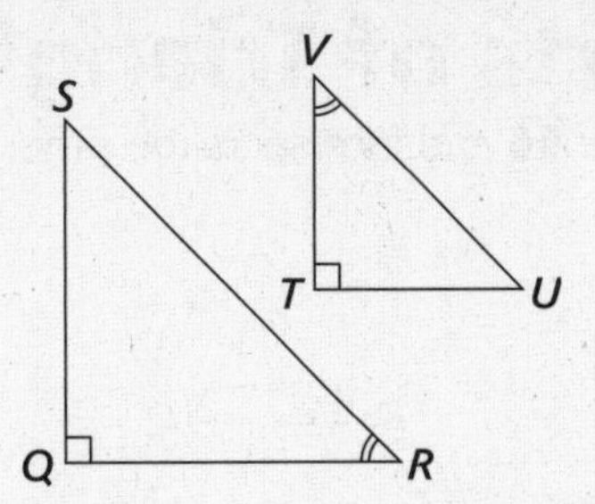

3. *Given*: $\triangle QRS$ and $\triangle TUV$, as marked in the preceding diagram. *Prove:* $\triangle QRS \sim \triangle TVU$.

Statement	***Reason***
1. $m\angle R = m\angle V$	1. Given (as marked).
2. $\angle Q = \angle T$ are right angles.	2. Given (as marked).
3. $m\angle Q = m\angle T$	3. All right angles are equal.
4. $\triangle QRS \sim \triangle TVU$	4. AA Similarity Postulate (18)

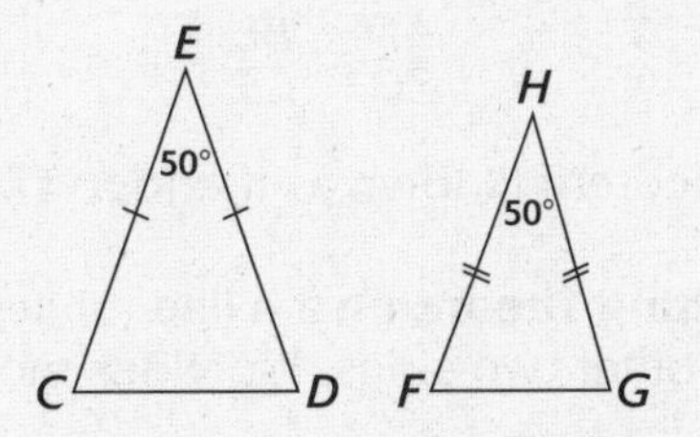

4. *Given*: $\triangle CDE$ and $\triangle FGH$, as marked in the preceding diagram. *Prove:* $\triangle CDE \sim \triangle FGH$.

Statement	***Reason***
1. $m\angle E = m\angle H$	1. Given (as marked).
2. $CE = DE$, $FH = GH$	2. Given (as marked).
3. $\triangle CDE$ and $\triangle FGH$ are isosceles triangles.	3. Definition of isosceles triangle.
4. $m\angle C = m\angle D$, $m\angle F = m\angle G$	4. Base angles of an isosceles triangle are equal or, in a triangle, angles opposite equal sides are equal.
5. $m\angle C = \frac{(180-50)}{2}$ $m\angle F = \frac{(180-50)}{2}$	5. Angle sum of any triangle = 180°.
6. $m\angle C = \frac{130}{2}$ $m\angle F = \frac{130}{2}$	6. Subtraction
7. $m\angle C = m\angle D = 65°$ $m\angle F = m\angle G = 65°$	7. Division and step 4
8. $m\angle C = m\angle D = m\angle F = m\angle G$	8. Angles equal to same measures
9. $\triangle CDE \sim \triangle FGH$	9. AA Similarity Postulate (18)

Proportional Parts of Triangles

In $\triangle ABC$, which follows, $\overline{DE} \parallel$ base $\overline{AB}$ and intersects the other two sides of the triangle at D and E.

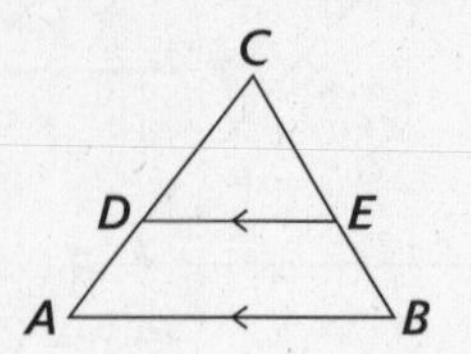

Using the AA Similarity Postulate, it is easy to prove that $\triangle ABC \sim \triangle DEC$. Then, since the corresponding sides of similar triangles are in proportion,

$$\frac{AC}{DC} = \frac{BC}{CE}$$

Next, we'll use the Denominator Subtraction Property to get the proportion:

$$\frac{AC - DC}{DC} = \frac{BC - CE}{CE}$$

However, you'll notice that $AC - DC = AD$ and $= BC - CE = BE$. That leads us to the following proportion:

$$\frac{AD}{CD} = \frac{BE}{CE}$$

We can conclude the following theorem (known as the [don't laugh!] Side-Splitting Theorem):

Theorem 54 (The Side-Splitting Theorem): If a line (or segment) is parallel to one side of a triangle and intersects the other two sides, it divides those sides proportionally.

More difficult to prove, but very useful in solving problems is the Angle Bisector Theorem. Check out the following diagram.

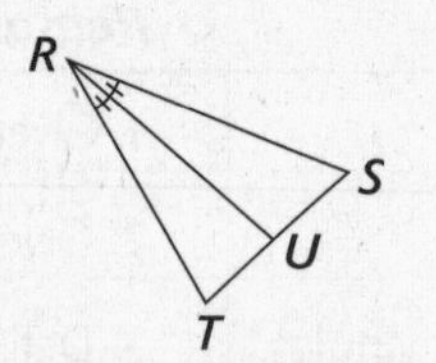

Theorem 55 (Angle Bisector Theorem): If a ray bisects an angle of a triangle, then it divides the opposite side into segments that are proportional to the sides forming the triangle.

That means $\frac{TU}{SU} = \frac{RT}{RS}$

or

$$\frac{TR}{TU} = \frac{RS}{SU}$$

Example Problems

These problems show the answers and solutions.

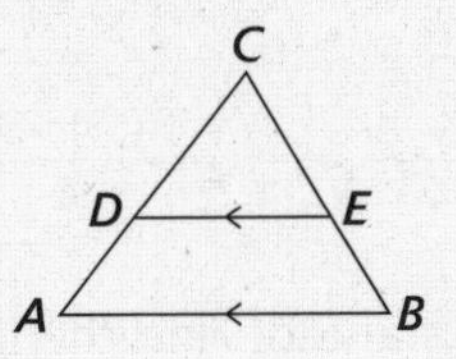

1. Suppose that $CE = 8$ cm, $BE = 6$ cm, and $CD = 12$ cm. To find x, the length of AD, write a proportion and solve it for x:

   ***Answer:* 9 cm** Here we solve by means of the Side-Splitting Theorem.

   Write a proportion: $\frac{AD}{CD} = \frac{BE}{CE}$

   Next, we substitute: $\frac{x}{12} = \frac{6}{8}$

   Use cross-products: $8x = 6 \cdot 12$

   Now multiply: $8x = 72$

   And divide by 8: $\frac{8x}{8} = \frac{72}{8}$

   To get: $x = 9$

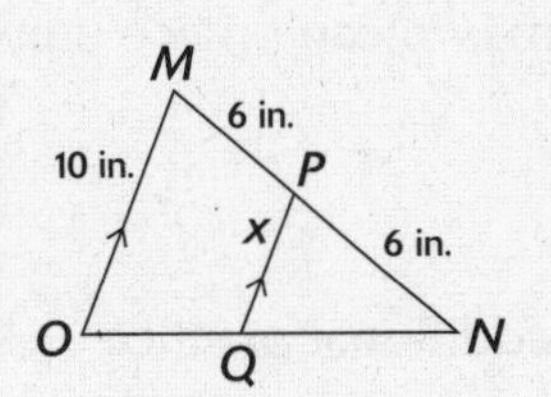

2. Find the length of $\overline{PQ}$ in the preceding diagram.

   ***Answer:* 5 in.** We can't use Theorem 54 this time, since PQ is not one of the sides being split. But, we first can find that $\triangle MNO \sim \triangle PNQ$. $\angle N$ is equal to itself by identity, and $m\angle M = m\angle NPQ$ (corresponding angles of parallel lines). You could have used the other pair of corresponding angles instead to get your AA similarity. Then, since the corresponding parts of similar triangles are in proportion,

   Write the proportion: $\frac{PQ}{MO} = \frac{NP}{MN}$

   Next, we substitute: $\frac{x}{10} = \frac{6}{12}$

   Use cross-products: $12x = 6 \cdot 10$

   Now multiply: $12x = 60$

   And divide by 12: $\frac{12x}{12} = \frac{60}{12}$

   To get: $x = 5$

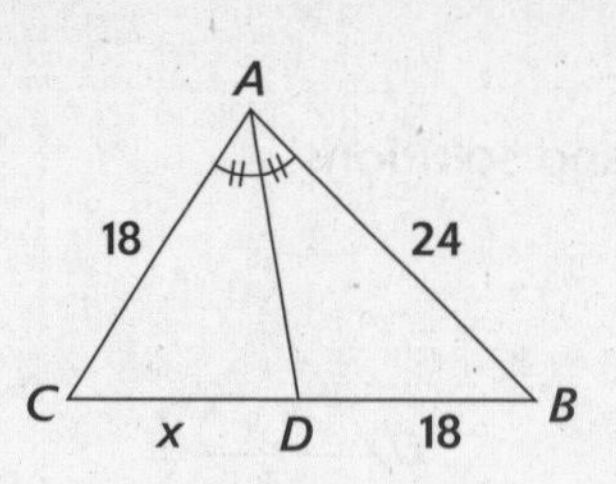

3. The dimensions of $\triangle ABC$ are as marked. Find the length of $\overline{CD}$.

***Answer:* 13.5** First, notice that $\angle A$ is bisected. Here is a perfect place for the Angle Bisector Theorem.

Write the proportion: $\frac{CD}{BD} = \frac{AC}{AB}$

Next, substitute values: $\frac{x}{18} = \frac{18}{24}$

Use cross-products: $24x = 18 \cdot 18$

Factor 6 out of each side to simplify: $4x = 3 \cdot 18$

Now multiply: $4x = 54$

And divide by 4: $\frac{4x}{4} = \frac{54}{4}$

To get: $x = 13.5$

Did you think they were all going to come out as whole numbers? Since units of measure were not shown on the diagram, we can't place units of measure in the answer.

Work Problems

Use these problems to give yourself additional practice.

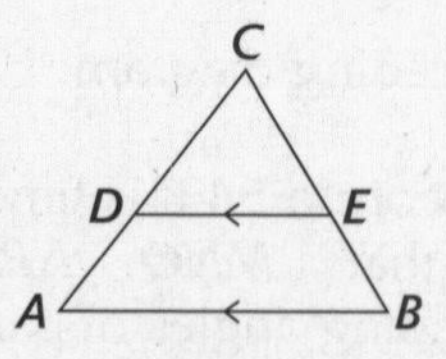

1. $CD = 8$ cm, $AD = 3$ cm, and $CE = 7$ cm. Find BE. Solve by using the Side-Splitting Theorem.

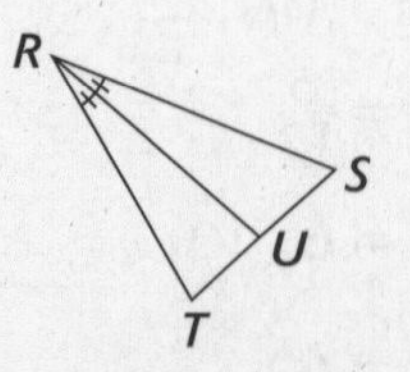

2. $RT = 16$ in., $TU = 4$ in., $SU = 5$ in. Find RS.

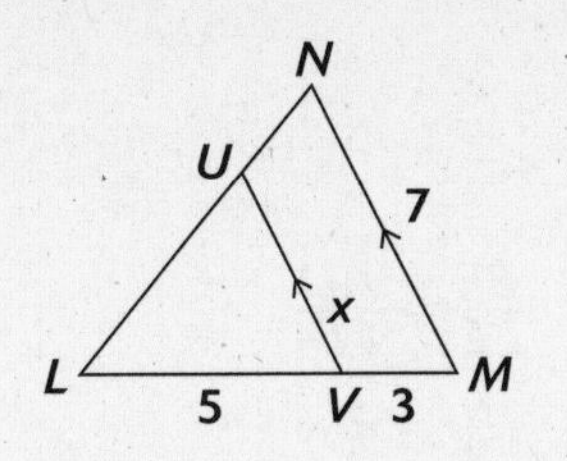

3. The dimensions and all other necessary information are marked on $\triangle LMN$. Find the length of $\overline{UV}$.

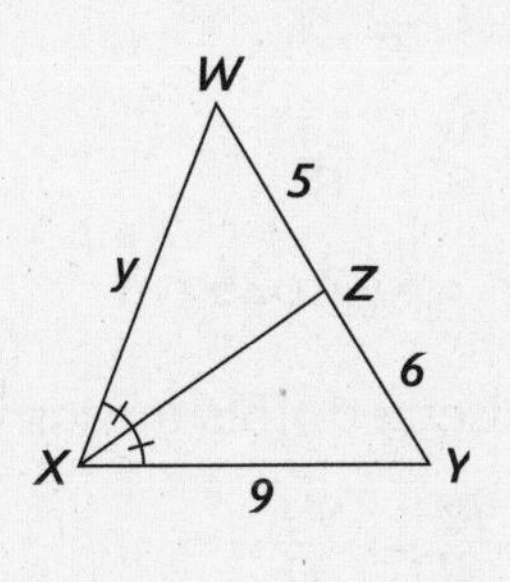

4. Find *WX*.

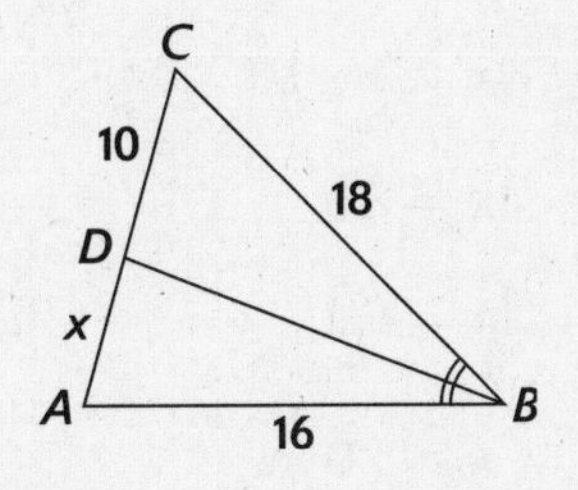

5. Find *AD*.

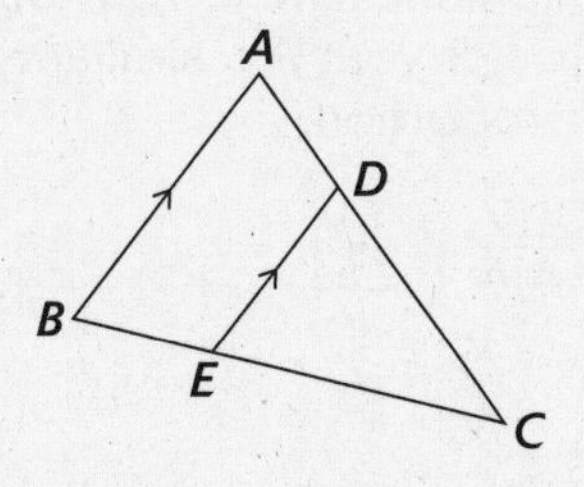

6. Suppose that $CE = 16$ cm, $BE = 12$ cm, and $CD = 24$ cm. Find the length of *AD* without proving the triangles similar.

Worked Solutions

1. **2.625 cm**

Write the proportion: $\frac{BE}{CE} = \frac{AD}{CD}$

Next, we substitute: $\frac{x}{7} = \frac{3}{8}$

Use cross-products: $8x = 3 \cdot 7$

Now multiply: $8x = 21$

And divide by 8: $\frac{8x}{8} = \frac{21}{8}$

To get: $x = 2.625$ cm

2. **20 in.** First notice that $\angle R$ is bisected. This means we use the Angle Bisector Theorem.

Write the proportion: $\frac{RS}{RT} = \frac{SU}{TU}$

Next, substitute values: $\frac{x}{16} = \frac{5}{4}$

Use cross-products: $4x = 5 \cdot 16$

Now multiply: $4x = 80$

And divide by 4: $\frac{4x}{4} = \frac{80}{4}$

To get: $x = 20$ in.

3. **4.375** We first must find that $\triangle LMN \sim \triangle LVU$. $\angle L$ is equal to itself by identity, and $\angle M = \angle LVU$ (corresponding angles of parallel lines). You could have used the other pair of corresponding angles instead to get your AA similarity. Then, since the corresponding parts of similar triangles are in proportion,

Write the proportion: $\frac{UV}{MN} = \frac{LV}{LM}$

Next, we substitute: $\frac{x}{7} = \frac{5}{8}$

Use cross-products: $8x = 5 \cdot 7$

Now multiply: $8x = 35$

And divide by 8: $\frac{8x}{8} = \frac{35}{8}$

To get: $x = 4.375$

4. **7.5** First notice that $\angle X$ is bisected. That means we use the Angle Bisector Theorem.

Write the proportion: $\frac{XW}{XY} = \frac{WZ}{YZ}$

Next, substitute values: $\frac{y}{9} = \frac{5}{6}$

Use cross-products: $6y = 5 \cdot 9$

Now multiply: $6y = 45$

And divide by 6: $\frac{6y}{6} = \frac{45}{6}$

To get: $y = 7.5$

5. $\mathbf{8\frac{8}{9}}$ You'll see that $\angle B$ is bisected. It's that Angle Bisector Theorem again.

Write the proportion: $\frac{AD}{AB} = \frac{CD}{BC}$

Next, substitute values: $\frac{x}{16} = \frac{10}{18}$

Use cross-products: $18x = 16 \cdot 10$

Now multiply: $18x = 160$

And divide by 18: $\frac{18x}{18} = \frac{160}{18}$

To get: $x = 8\frac{8}{9}$

6. **18 cm** Here we solve by means of the Side-Splitting Theorem.

Write the proportion: $\frac{AD}{CD} = \frac{BE}{CE}$

Next, we substitute: $\frac{x}{24} = \frac{12}{16}$

Use cross-products: $16x = 24 \cdot 12$

Now multiply: $16x = 288$

And divide by 16: $\frac{16x}{16} = \frac{288}{16}$

To get: $x = 18$ cm

Proportional Parts of Similar Triangles

We already have studied the meaning of angle bisectors and altitudes. A median of a triangle is a line segment that begins at a vertex and bisects the side opposite that vertex. That brings us to Theorem 56.

> **Theorem 56:** If two triangles are similar, then the ratio of any two corresponding segments in those triangles (altitudes, medians, and angle bisectors) are equal to the ratio between any two corresponding sides.

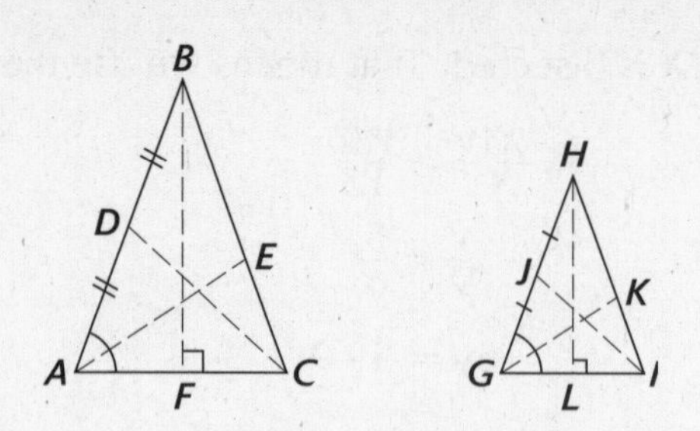

In the preceding figure, $\triangle ABC \sim \triangle GHI$. Therefore, $\frac{AB}{GH} = \frac{BC}{HI} = \frac{AC}{GI}$ because corresponding sides of similar triangles are in the same ratio (are proportional).

According to Theorem 56,

$$\frac{\text{length of angle bisector } AE}{\text{length of angle bisector } GK} = \frac{AB}{GH}$$

$$\frac{\text{length of altitude } BF}{\text{length of altitude } HL} = \frac{AB}{GH}$$

$$\frac{\text{length of median } CD}{\text{length of median } IJ} = \frac{AB}{GH}$$

Example Problems

These problems show the answers and solutions.

1. Two similar triangles have sides in the ratio 3:4. The median of the first triangle is 12 inches long. How long is the median of the second triangle?

   ***Answer:* 16 in.** The ratios of median to median and side to side must be the same according to Theorem 56. So,

Build the proportion:	$\frac{12}{x} = \frac{3}{4}$
Cross multiply:	$3x = 4 \cdot 12$
Multiply:	$3x = 48$
Divide in your head:	$x = 16$ in.

2. $\triangle QRS \sim \triangle MNO$. The shortest side of $\triangle QRS$ is 14 ft. long, and its longest side is 24. The shortest side of $\triangle MNO$ is 21 ft. long. An altitude of the larger triangle is 27 ft. long. What is the length of the corresponding altitude of $\triangle QRS$?

   ***Answer:* 18 ft.** The ratios of altitude to altitude and side to side must be the same according to Theorem 56. According to the problem, the ratio of the shortest sides is 14:21. By factoring out a 7 (dividing both numbers by that amount), that simplifies to 2:3. Remember to build the proportion ordering smaller to larger.

Build the proportion: $\frac{x}{27} = \frac{2}{3}$

Cross multiply: $3x = 2 \cdot 27$

Multiply: $3x = 54$

Divide in your head: $x = 18$ ft.

3. $\triangle LMN \sim \triangle TUV$. The median from $\angle L$ to $\overline{MN}$ is 8 cm long. The median from $\angle T$ to $\overline{UV}$ is 40 cm long. Side $\overline{LM}$ of the first triangle is 9 cm long. Its corresponding side in the second triangle is $\overline{TU}$. How long is TU?

***Answer:* 45 cm** The ratios of angle bisector to angle bisector and side to side must be the same according to Theorem 56. The two medians are in the ratio 8:40. Don't look now, but we can factor an 8 out of that ratio. So,

First the proportion: $\frac{9}{x} = \frac{1}{5}$

Cross multiply: $x = 5 \cdot 9$

Multiply: $x = 45$ cm

Perimeter and Areas of Similar Triangles

When two triangles are similar, a ratio exists between each pair of corresponding sides. When that ratio (a.k.a. fraction) is expressed in lowest terms, it is known as the **scale factor** of the two triangles. Here's an example:

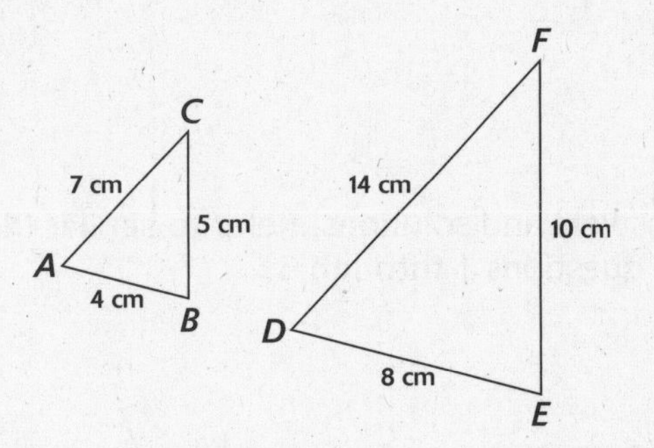

$\triangle ABC \sim \triangle DEF$. The ratios of the corresponding sides are $\frac{7}{14}$, $\frac{5}{10}$, and $\frac{4}{8}$. Each of those fractions simplifies to $\frac{1}{2}$. That means the scale factor for these two triangles is 1:2.

The perimeter of $\triangle ABC$ is found by adding 7 cm + 5 cm + 4 cm to get 16 cm. To find the perimeter of $\triangle DEF$, add 14 cm + 10 cm + 8 cm to get 32 cm. Comparing them shows that the ratio of the triangles' perimeters is 16:32, which simplifies to 1:2. Hmm! That sounds familiar. In fact, it leads to the following theorem:

Theorem 57: If two similar triangles have a scale factor of a:b, then the ratio of their perimeters is a:b.

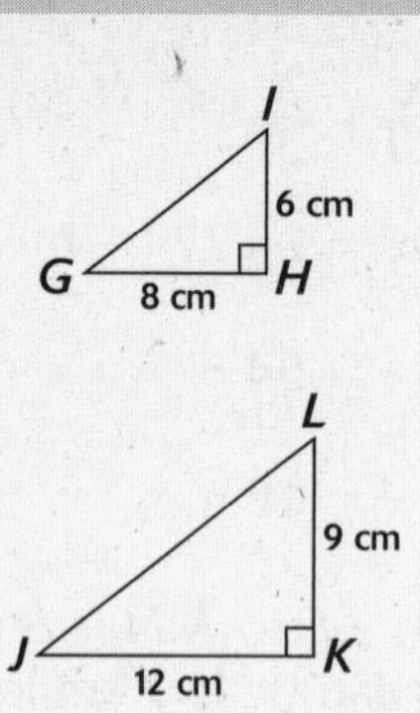

The preceding two right triangles have a scale factor of 2:3. Let's find their areas:

$$\text{Area}\triangle GHI = \frac{1}{2}(6)(8) \qquad \text{Area}\triangle JKL = \frac{1}{2}(9)(12)$$

$$\text{Area}\triangle GHI = 24\text{ cm}^2 \qquad \text{Area}\triangle JKL = 54\text{ cm}^2$$

Next, let's compare the areas of these two similar triangles by making a ratio:

$$\frac{\text{Area}\triangle GHI}{\text{Area}\triangle JKL} = \frac{24}{54} = \frac{4}{9} = \left(\frac{2}{3}\right)^2$$

From this relationship, we get the following theorem relating areas.

Theorem 58: If two similar triangles have a scale factor of $a{:}b$, then the ratio of their areas is $a^2{:}b^2$.

Example Problems

These problems show the answers and solutions. Refer to similar triangles $\triangle ABC$ and $\triangle DEF$ in the following figure to solve questions 1 through 3.

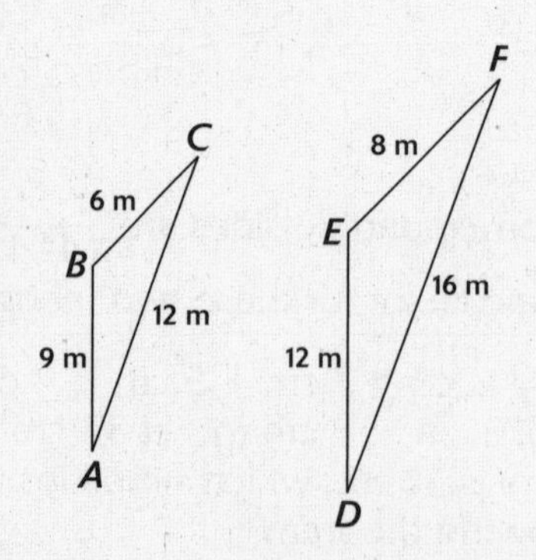

1. What is the scale factor of $\triangle DEF$ to $\triangle ABC$?

   ***Answer:* 4:3** Crucial is the fact that the question placed the triangles in the order of larger to smaller. That means the comparisons must be made in that same order: $\frac{8}{6} = \frac{16}{12} = \frac{12}{9}$. Each simplifies to $\frac{4}{3}$, or a ratio of 4:3.

2. What is the ratio of the perimeter of $\triangle ABC$ to $\triangle DEF$?

Answer: 3:4 Theorem 57 tells us: If two similar triangles have a scale factor of *a*:*b*, then the ratio of their perimeters is *a*:*b*. That would make the ratio of the perimeters of $\triangle DEF$ to $\triangle ABC$ 4:3. For the ratio of the perimeters of $\triangle ABC$ to $\triangle DEF$ that ratio is reversed to 3:4.

3. What is the ratio of the areas of $\triangle ABC$ to $\triangle DEF$?

Answer: 9:16 Theorem 58 tells us: If two similar triangles have a scale factor of *a*:*b*, then the ratio of their areas is $a^2:b^2$. Since the scale factor of $\triangle ABC$ to $\triangle DEF$ is 3:4, the ratio of the areas is $3^2:4^2$. That's 9:16.

Work Problems

Use these problems to give yourself additional practice. Refer to similar triangles $\triangle ABC$ and $\triangle DEF$ in the following figure to answer questions 1–6.

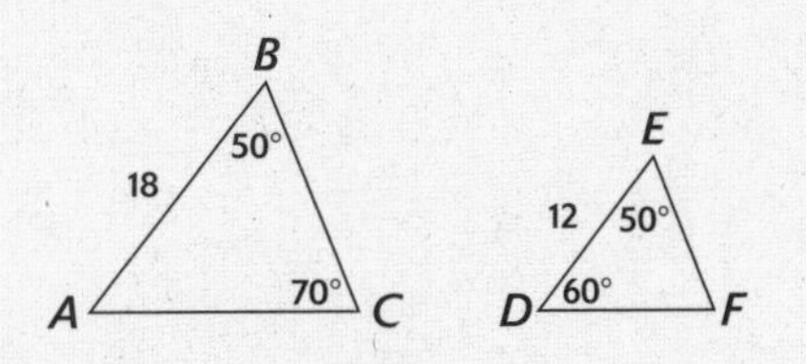

1. What is the scale factor of $\triangle ABC$ to $\triangle DEF$?
2. What is the scale factor of $\triangle DEF$ to $\triangle ABC$?
3. What is the ratio of the areas of $\triangle DEF$ to $\triangle ABC$?
4. What is the ratio of the areas of $\triangle ABC$ to $\triangle DEF$?
5. What is the ratio of the perimeters of $\triangle DEF$ to $\triangle ABC$?
6. What is the ratio of the perimeters of $\triangle ABC$ to $\triangle DEF$?

Worked Solutions

1. **3:2** $\frac{18}{12}=\frac{15}{10}=\frac{3}{2}$. The scale factor is the ratio of one to another *in the order mentioned*, expressed *in lowest terms*.

2. **2:3** $\frac{12}{18}=\frac{10}{15}=\frac{2}{3}$. The scale factor is the ratio of one to another *in the order mentioned*, expressed in *lowest terms*.

3. **4:9** The ratio of the areas equals the squares of the ratio of the sides (Theorem 58).

$$\left(\frac{2}{3}\right)^2 = \frac{4}{9}$$

4. **9:4** The ratio of the areas equals the squares of the ratio of the sides (Theorem 58).

$$\left(\frac{3}{2}\right)^2 = \frac{9}{4}$$

5. **2:3** The ratio of the perimeters equals the scale factor of the sides (Theorem 57).

6. **3:2** The ratio of the perimeters equals the scale factor of the sides (Theorem 57). Remember to pay attention to order.

Chapter 7

Right Triangles

Right triangles are pretty special, as you've already seen from the fact that they have their own special congruence postulates. They are, in fact, so special that they command their own branch of mathematics—trigonometry. Although that's way beyond the scope of this book, we will deal with two of the very special triangles that all trigonometry students learn about: the 45°-45°-90° triangle, and the 30°-60°-90° triangle. We'll also study the Pythagorean Theorem, which led to the Pythagorean Identities, upon which trigonometry is based.

Geometric Mean

When a positive value is repeated in the means or extremes of a proportion, it is referred to as the **mean proportional**, or the **geometric mean** between the other two numbers. You'll see how it is used in the example problems that follow.

Example Problems

These problems show the answers and solutions.

1. Find the mean proportional between 4 and 16.

   ***Answer:* 8** Let x represent the mean proportional.

   $\frac{4}{x} = \frac{x}{16}$ (definition of mean proportional)

   $x^2 = 64$ (by cross-products)

   $x = \sqrt{64}$ $\quad x = 8$

   The mean proportional between 4 and 16 is 8.

2. 10 is the geometric mean between 4 and what other number?

   ***Answer:* 25** Let x represent the other number.

   $\frac{4}{10} = \frac{10}{x}$ (definition of geometric mean)

   $4x = 100$ (by cross-products)

   $x = 25$

   10 is the geometric mean between 4 and 25.

Altitude to the Hypotenuse

In the diagram that follows, right triangle ABC is drawn with $\overline{BD}$ drawn as an altitude to $\overline{AC}$.

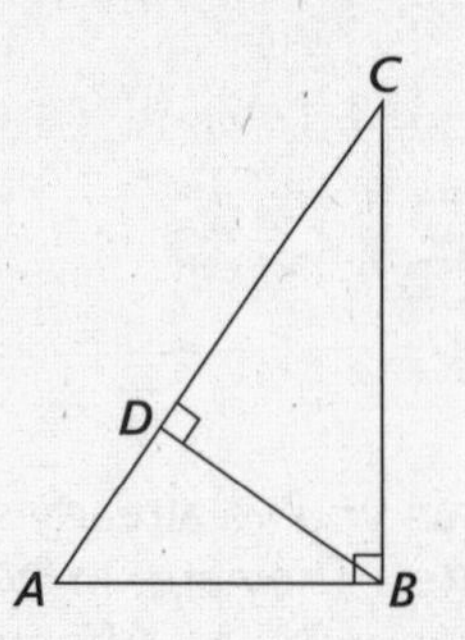

Using the AA similarity postulate, it is quite easy to prove the following theorem:

Theorem 59: If the altitude is drawn to the hypotenuse of a right triangle, it creates two similar right triangles, each similar to the original triangle as well as to each other.

In $\triangle ABD$, $\overline{AB}$ is the hypotenuse, corresponding to $\overline{AC}$ in the big triangle and $\overline{BC}$ in $\triangle BDC$. Check out the figure that follows.

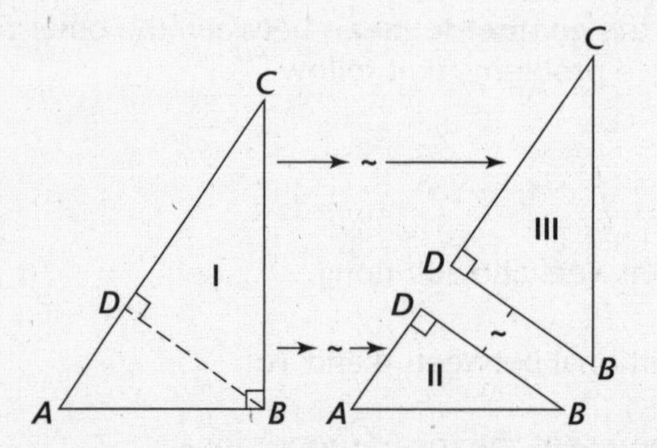

Notice which other parts of the three similar triangles correspond to each other. $\overline{AB}$ in $\triangle$I corresponds to $\overline{AD}$ in $\triangle$II and to $\overline{BD}$ in $\triangle$III. Also, $\overline{BC}$ in $\triangle$I corresponds to $\overline{BD}$ in $\triangle$II and to $\overline{CD}$ in $\triangle$III. Since corresponding sides of similar triangles are in proportion, we have three proportions created that involve geometric means:

First, $\dfrac{AC}{BC} = \dfrac{BC}{CD}$ $(\triangle\text{I} \sim \triangle\text{III})$

Second, $\dfrac{AD}{AB} = \dfrac{AB}{CD}$ $(\triangle\text{II} \sim \triangle\text{I})$

These two proportions may be summarized in the following theorem.

Theorem 60: If the altitude is drawn to the hypotenuse of a right triangle, then each leg is the geometric mean between the hypotenuse and the segment on the hypotenuse that it touches.

There's still one more proportion to be examined.

Third, $\dfrac{AD}{BD} = \dfrac{BD}{CD}$ $(\triangle\text{III} \sim \triangle\text{II})$

This proportion can be stated as the following theorem:

> **Theorem 61:** If an altitude is drawn to the hypotenuse of a right triangle, then it is the geometric mean between the segments of the hypotenuse.

Example Problems

These problems show the answers and solutions. Using Theorems 60 and 61, write three separate proportions using geometric means based upon the figure that follows.

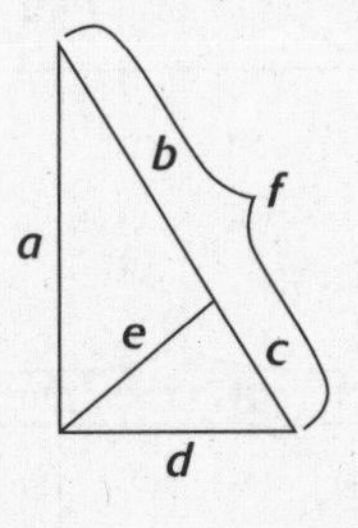

1. Use *a* and *f*.

   Answer: $\frac{f}{a} = \frac{a}{b}$

   By Theorem 60, the leg, *a*, is the mean proportional (or geometric mean) between the hypotenuse and the segment of the hypotenuse that the leg touches, *b*.

2. Use *d* and *f*.

   Answer: $\frac{f}{d} = \frac{d}{c}$

   By Theorem 60, the leg, *d*, is the mean proportional (or geometric mean) between the hypotenuse and the segment of the hypotenuse that the leg touches, *c*.

3. Use *e* and *b* and *c*.

   Answer: $\frac{b}{e} = \frac{e}{c}$

   By Theorem 61, the altitude to the hypotenuse, *e*, is the geometric mean for the segments of the hypotenuse, *b* and *c*.

Work Problems

Use these problems to give yourself additional practice.

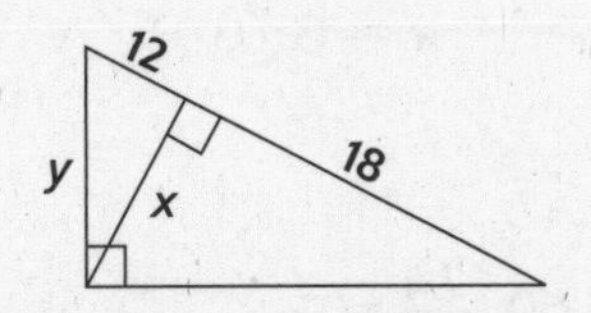

Use the preceding figure to answer questions 1 and 2.

1. Find the length of *y*.

2. Find the length of *x*.

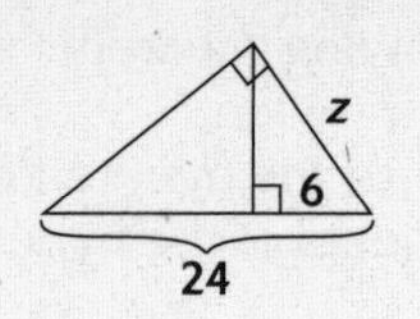

3. Find the length of z.

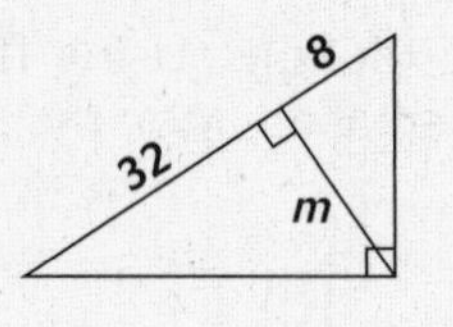

4. Find the length of m.

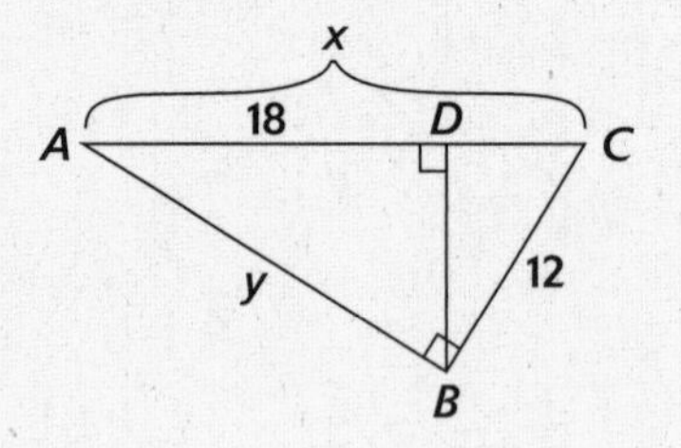

5. Find the lengths of x and y.

Worked Solutions

1. **$6\sqrt{10}$**

By Theorem 60, $\frac{30}{y} = \frac{y}{12}$

$$y^2 = 360 \quad \text{(by cross-products)}$$
$$y = \sqrt{360}$$
$$y = \left(\sqrt{36}\right)\left(\sqrt{10}\right)$$
$$y = 6\sqrt{10}$$

2. **$6\sqrt{6}$**

By Theorem 61, $\frac{12}{x} = \frac{x}{18}$

$$x^2 = 216 \quad \text{(by cross-products)}$$
$$x = \sqrt{216}$$
$$x = \left(\sqrt{36}\right)\left(\sqrt{6}\right)$$
$$x = 6\sqrt{6}$$

3. **12**

By Theorem 60, $\frac{24}{z} = \frac{z}{6}$

$$z^2 = 144 \quad \text{(by cross-products)}$$
$$z = \sqrt{144}$$
$$z = 12$$

4. **16**

$$\text{By Theorem 61, } \frac{32}{m} = \frac{m}{8}$$
$$m^2 = 256 \quad \text{(by cross-products)}$$
$$m = \sqrt{256}$$
$$m = 16$$

5. $\mathbf{x = 24, y = 12\sqrt{3}}$

$$AD + CD = AC \quad \text{(by the Segment Addition Postulate)}$$
$$CD + 18 = x$$
$$CD = x - 18$$
$$\text{By Theorem 60, } \frac{x}{12} = \frac{12}{(x-18)}$$
$$x(x - 18) = 144 \quad \text{(by cross-products)}$$
$$x^2 - 18x = 144$$
$$x^2 - 18x - 144 = 0$$
$$\text{Factor: } (x - 24)(x + 6) = 0$$
$$x - 24 = 0 \text{ or } x + 6 = 0$$
$$x = 24, \text{ or } x = -6$$

Since it stands for a length of a side, x cannot be negative, so $x = 24$.

$$\text{By Theorem 60, } \frac{24}{y} = \frac{y}{18}$$
$$y^2 = 432 \quad \text{(by cross-products)}$$
$$y = \sqrt{432}$$
$$y = (\sqrt{144})(\sqrt{3})$$
$$y = 12\sqrt{3}$$

The Pythagorean Theorem

We have no interest in deriving the Pythagorean Theorem. It is usually stated as follows: in any right triangle, the square of the hypotenuse is equal to the sum of the squares of the legs. Although this is mathematically correct, it does not really give you an understanding of what it was that Pythagoras actually discovered over MM (2000) years ago. To get a better appreciation of it, look at the diagram that follows.

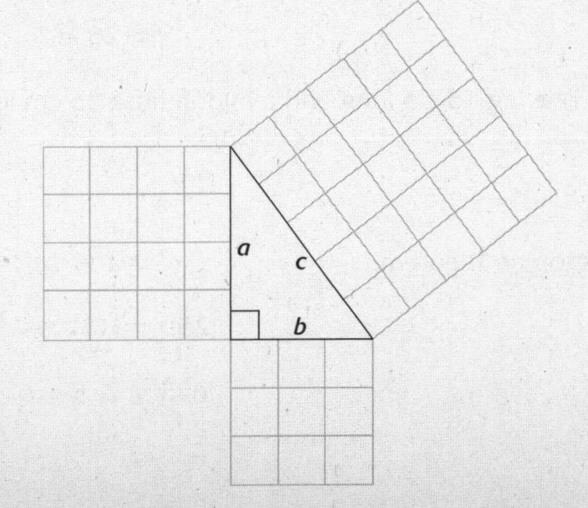

With that figure in mind, consider this:

> **Theorem 62** (The Pythagorean Theorem): The square ***on*** the hypotenuse of a right triangle *is equal in area* to the ***sum*** *of the areas* of the squares on the two legs.

If we call the legs of a right triangle *a* and *b*, and the hypotenuse *c*, then $c^2 = a^2 + b^2$.

Example Problems

These problems show the answers and solutions.

1. A right triangle has legs 9 cm long and 12 cm long. How long is its hypotenuse?

   ***Answer:* 15 cm**

   Start with the Pythagorean Theorem: $c^2 = a^2 + b^2$

   Substitute values: $c^2 = (9)^2 + (12)^2$

   Square the values: $c^2 = 81 + 144$

   Add: $c^2 = 225$

   Take the square root of both sides: $\sqrt{c^2} = \sqrt{225}$

   Find the answer: $c = 15$ cm

2. A right triangle has legs 10 in. long and 24 in. long. How long is its hypotenuse?

   ***Answer:* 26 in.**

   Start with the Pythagorean Theorem: $c^2 = a^2 + b^2$

   Substitute values: $c^2 = (10)^2 + (24)^2$

   Square the values: $c^2 = 100 + 576$

   Add: $c^2 = 676$

   Take the square root of both sides: $\sqrt{c^2} = \sqrt{676}$

   Find the answer: $c = 26$ in.

3. A right triangle has one leg 15 cm long and a hypotenuse 25 cm long. How long is the other leg?

   ***Answer:* 20 cm**

   Start with the Pythagorean Theorem: $c^2 = a^2 + b^2$

   Substitute values: $(25)^2 = (15)^2 + b^2$

   Square the values: $625 = 225 + b^2$

Subtract 225 from both sides:	$625 - 225 = 225 + b^2 - 225$
Reverse the equation for convenience:	$b^2 = 400$
Take the square root of both sides:	$\sqrt{b^2} = \sqrt{400}$
Find the answer:	$c = 20$ cm

Pythagorean Triples

The Pythagorean formula $c^2 = a^2 + b^2$ may also be stated as $a^2 + b^2 = c^2$. With the numbers in alphabetical order, consider any combination of whole numbers, a, b, and c, that make the mathematical sentence $a^2 + b^2 = c^2$ true. Those three numbers are called a Pythagorean triple. 3-4-5 is a Pythagorean triple, since substituting those numbers for a, b, and c makes $a^2 + b^2 = c^2$ true.

$$(3)^2 + (4)^2 = (5)^2$$
$$9 + 16 = 25$$

Those numbers should look familiar, since they were the lengths of the sides of the triangle in the drawing at the opening of the last section. They are also the numbers that (times 3) were used in example 1, (9-12-15) and (times 5) were used in example 3 (15-20-25). 5-12-13 and 8-15-17 are also common Pythagorean triples. Memorize them and look for them when you have to solve problems involving Theorem 62. You may save a lot of time that way, and on tests, saving time on one type of problem means that you'll have more to use on others.

One other interesting use of the Pythagorean Theorem is its inverse, Theorem 63, which gives us a way to test whether a certain triangle is or is not a right triangle.

Theorem 63: If a triangle has sides of lengths a, b, and c, where c is the longest side, and $c^2 = a^2 + b^2$, then that triangle is a right triangle, and c is its hypotenuse.

Example Problems

These problems show the answers and solutions. Solve each of these problems mentally, by finding the Pythagorean triple.

1. A right triangle has legs 24 in. long and 18 in. long. How long is its hypotenuse?

   ***Answer:* 30 in.** The Pythagorean triple used is 3-4-5 (times 6). $6 \times 3 = 18$, $6 \times 4 = 24$, and $6 \times 5 = 30$.

2. A right triangle has legs 15 mm long and 36 mm long. How long is its hypotenuse?

   ***Answer:* 39 mm** The Pythagorean triple used is 5-12-13 (times 3). $3 \times 5 = 15$, $3 \times 12 = 36$, and $3 \times 13 = 39$.

3. A right triangle has one leg 32 cm long and a hypotenuse 68 cm long. How long is the other leg?

   ***Answer:* 60 cm** The Pythagorean triple used is 8-15-17 (times 4). $8 \times 4 = 32$, $15 \times 4 = 60$, and $17 \times 4 = 68$.

Work Problems

Use these problems to give yourself additional practice.

1. A 20 ft. long ladder leans against the side of a house. The bottom of the ladder is 12 ft. from the house. How far up from the ground does the ladder touch the house?
2. The diagonal distance across a rectangular parking lot is 65 yards. The shorter side of the parking lot is 25 yards. What is the parking lot's area?
3. A triangle has sides of lengths 16-34-30 inches. Is it a right triangle?
4. A triangle has sides of lengths 12-15-18 ft. Is it a right triangle?
5. A triangle has sides of lengths 7-13-9 cm. Is it a right triangle?

Worked Solutions

1. **16 ft.** The ladder is the hypotenuse of a right triangle.

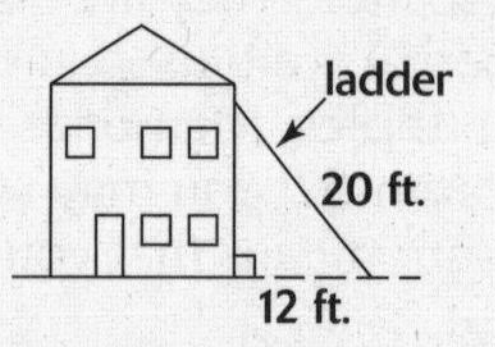

The height up the house forms one leg, and the distance along the ground between the base of the ladder and the house is the other leg. 12 and 20 are the 3-leg and hypotenuse of a 3-4-5 (times 4) right triangle. The 4-leg is missing.

$4 \times 4 = 16$ ft.

2. **1500 yd²**

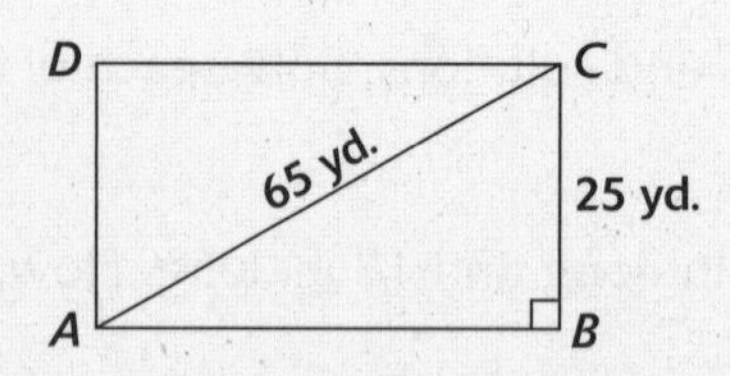

The diagonal of the rectangular parking lot forms the right triangles *ABC* and *ACD*. In either case, the missing side can be found as follows:

Start with the Pythagorean Theorem: $c^2 = a^2 + b^2$

Substitute values: $(65)^2 = (25)^2 + b^2$

Square the values: $4225 = 625 + b^2$

Subtract 625 from both sides: $4225 - 625 = 625 + b^2 - 625$

Reverse the equation for convenience:	$b^2 = 3600$
Take the square root of both sides:	$\sqrt{b^2} = \sqrt{3600}$
Find the leg:	$c = 60$ yd.

Or, you could have recognized the Pythagorean triple, 5-12-13 (× 5). Now that we have the adjacent sides of the rectangular parking lot, we find the area by

$$A = l \times w$$
$$A = 25 \times 60$$
$$A = 1500 \text{ yds}^2$$

3. Yes

Test the numbers:	$c^2 = a^2 + b^2$
The longest must be *c* (Theorem 63).	$(34)^2 = (16)^2 + (30)^2$
	$1156 = 256 + 900$
	$1156 = 1156$

That works! By the way, it was 8-15-17 (× 2).

4. No

Test the numbers:	$c^2 = a^2 + b^2$
The longest must be *c* (Theorem 63):	$(18)^2 = (12)^2 + (15)^2$
	$324 = 144 + 225$
	$324 = 369$ is not a true statement, so a triangle with sides of lengths 12-15-18 cm is *not* a right triangle.

5. No

Test the numbers:	$c^2 = a^2 + b^2$
The longest must be *c* (Theorem 63):	$(13)^2 = (7)^2 + (9)^2$
	$169 = 49 + 81$
	$169 = 130$ is not a true statement, so a triangle with sides of lengths 7-13-9 cm is *not* a right triangle.

Outgrowths of the Pythagorean Theorem

Theorem 63 can lead to conclusions about all triangles, based upon the relationships of the lengths of their sides. Specifically, from an examination of the relationships of the squares of the sides, we can determine whether a triangle is right (from Theorem 63), but we can also find out whether a triangle is acute or obtuse, from the following:

> **Theorem 64:** If a triangle has sides of lengths *a*, *b*, and *c*, where *c* is the longest length, and $c^2 > a^2 + b^2$, then that triangle is an obtuse triangle.
>
> **Theorem 65:** If a triangle has sides of lengths *a*, *b*, and *c* where *c* is the longest length, and $c^2 < a^2 + b^2$, then that triangle is an acute triangle.

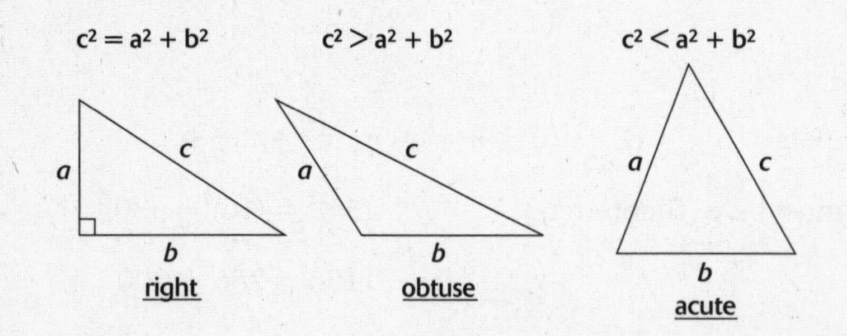

The preceding figure illustrates Theorems 63 through 65. Notice in the case of the middle triangle, side *c* is not larger than $a + b$; rather, *c*'s square is larger than the sum of the other sides' squares.

Example Problems

These problems show the answers and solutions. For each of the following sets of side-lengths, first decide whether the values could be a triangle. Then, if it is, tell whether the triangle is right, acute, or obtuse. Note that a "?" is used in each mathematical sentence instead of a "=", "<", or ">". Treat it as if it were one of those relaters.

1. **10-10-16**

   Answer: **obtuse**

   $c^2 ? a^2 + b^2$

   Substitute: $(16)^2 ? (10)^2 + (10)^2$

   Square each quantity: $256 ? 100 + 100$

   Add: $256 ? 200$

   Relate: $256 > 200$, so the triangle is obtuse (Theorem 64).

2. **12-13-16**

   Answer: **acute**

   $c^2 ? a^2 + b^2$

   $(16)^2 ? (13)^2 + (12)^2$

Square each quantity: 256 ? 169 + 144

Add: 256 ? 313

Relate: 256 < 313, so the triangle is acute (Theorem 65).

3. 24-45-51

***Answer:* right**

$c^2 ? a^2 + b^2$

$(51)^2 ? (24)^2 + (45)^2$

Square each quantity: 2601 ? 576 + 2025

Add: 2601 ? 2601

Relate: 2601 = 2601, so the triangle is right (Theorem 63).

Special Right Triangles

As if right triangles were not special enough, even more special types of right triangles exist. An equilateral right triangle is not one of them; if you think about it, you'll realize that it's an impossibility.

Isosceles Right Triangle

An **isosceles right triangle** has the properties of both a right triangle and an isosceles triangle. It has two equal sides, two equal angles (opposite those equal sides or legs), and one right angle. Therefore, an isosceles triangle will always look like the ones in the following figure.

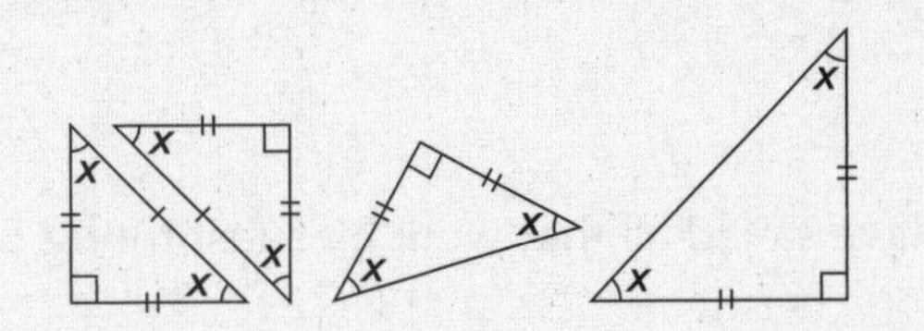

The orientation of the triangle (which way it faces) is meaningless. In all cases, both sides are the same length, and all nonright angles are the same size. To find that angle, use the fact that the angles of any triangle must sum to 180°. Remember, we start out with one 90° angle and two other equal ones, so

$$x + x + 90° = 180°$$
$$2x + 90° = 180°$$
$$2x = 90°$$
$$x = 45°$$

That's the only nonright angle you'll ever find in an isosceles right triangle.

The sides of an isosceles right triangle are always in the same ratio, namely $1:1:\sqrt{2}$. We can easily prove that using the Pythagorean Theorem, but we'll leave it for you to do that if you're skeptical. That ratio can also be written as it is in the figure that follows.

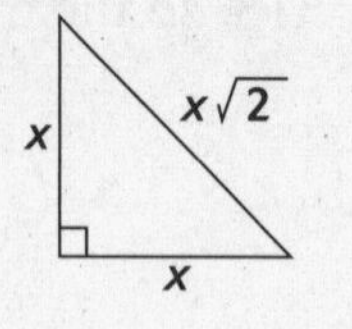

Example Problems

These problems show the answers and solutions.

1. An isosceles right triangle has one leg that is 5 inches long. How long are the other two sides?

   ***Answer:* 5 inches, $5\sqrt{2}$ inches** Since one leg is 5 inches, and the triangle is isosceles, the other leg must be 5 inches. Then, to get the hypotenuse, use the Pythagorean Theorem.

   $$c^2 = a^2 + b^2$$

   $$c^2 = (5)^2 + (5)^2$$

   Square each quantity: $c^2 = 25 + 25$

   Add: $c^2 = 50$

   Take the square root of both sides: $\sqrt{c^2} = \sqrt{50}$

   $\sqrt{50}$ can be simplified: $c = \sqrt{25 \cdot 2}$

   Finally, we get: $c = 5\sqrt{2}$ in.

   Of course, we could have saved all that work by substituting 5 for the x in the diagram.

2. An isosceles right triangle has one leg that is 9 cm long. How long are the other two sides?

   ***Answer:* 9 cm, $9\sqrt{2}$ cm** This time, we'll learn from our mistake in the last problem. Substituting 9 for the x on the figure, we find that the other leg is 9 cm, and the hypotenuse is $9\sqrt{2}$ cm.

3. The hypotenuse of a right isosceles triangle is 10 in. What is the length of each leg?

   ***Answer:* $5\sqrt{2}$ in.** Hmm. If that hypotenuse had a $\sqrt{2}$ attached to it, you would have probably known what to do, but, since it does not, it's back to Pythagoras:

$$c^2 = a^2 + b^2$$

$$(10)^2 = (x)^2 + (x)^2$$

Combine quantities and reverse: $2x^2 = 100$

Add: $x^2 = 50$

Take the square root of both sides: $\sqrt{x^2} = \sqrt{50}$

$\sqrt{50}$ can be simplified: $x = \sqrt{25 \cdot 2}$

Finally, we get: $x = 5\sqrt{2}$ in.

30-60-90 Right Triangle

A right triangle containing one 30° and one 60° angle is another special right triangle, known as a **30°-60°-90° right triangle**, or, more commonly, a **30-60-90 triangle**. The ratios of the sides of a 30-60-90 triangle are shown in the following figure.

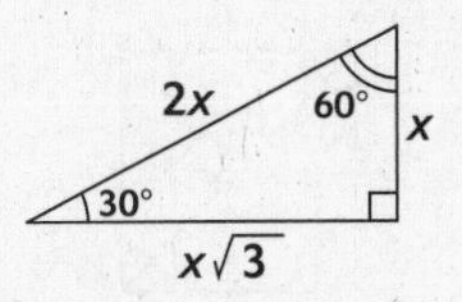

The side opposite the 30° angle is half the hypotenuse, and the side opposite the 60° angle is half the hypotenuse times $\sqrt{3}$.

Example Problems

These problems show the answers and solutions.

1. A 30-60-90 triangle's shortest side is 11 inches long. What is the length of the triangle's second shortest side?

   ***Answer:* $11\sqrt{3}$** The sides of a 30-60-90 triangle are, in order of sides opposite each angle, half the hypotenuse, half the hypotenuse times $\sqrt{3}$, and twice the shortest side.

2. The hypotenuse of a 30-60-90 right triangle is $8\sqrt{3}$. What are the lengths of the other two sides?

   ***Answer:* $4\sqrt{3}$ and 12** The side opposite the 30° angle is half the hypotenuse. That's $\frac{1}{2}(8\sqrt{3})$, which works out to $4\sqrt{3}$. The side opposite the 60° angle is half the hypotenuse times $\sqrt{3}$. That's $4(\sqrt{3})(\sqrt{3})$. But $(\sqrt{3})(\sqrt{3}) = 3$, so the side is $4 \times 3 = 12$.

3. Find the altitude of an equilateral triangle with a perimeter of 18.

   ***Answer:* $3\sqrt{3}$** An equilateral triangle with a perimeter of 18 has a side that's one-third the perimeter, or 6. Draw any altitude of an equilateral triangle, and you form two 30-60-90 triangles with the altitude being the side opposite both 60° angles. That means its length is half the hypotenuse (now either of the original triangle's sides) times $\sqrt{3}$. Half of 6 makes 3 times $\sqrt{3} = 3\sqrt{3}$.

Work Problems

Use these problems to give yourself additional practice.

1. Find the perimeter of a square whose diagonal is $5\sqrt{2}$ cm.
2. The shortest side of a 30-60-90 triangle is 7 ft. How long is the hypotenuse?
3. The second longest side of a 30-60-90 triangle is $9\sqrt{3}$ m. How long is the hypotenuse?
4. Find the diagonal of a square whose side is 7 cm.
5. What is the area of an equilateral triangle whose side is 12 in?

Worked Solutions

1. **20 cm** This could be done by the Pythagorean Theorem, but why bother? The diagonal of a square cuts the figure into two isosceles right triangles. Remember, the diagonal of a square bisects the 90° corner angles, making two 45°-45°-90° triangles.

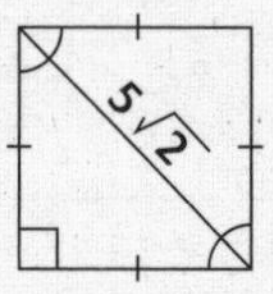

The diagonal of the square is the hypotenuse of both right triangles. Following the pattern we learned earlier, in a right isosceles triangle with side = x, the hypotenuse is $x\sqrt{2}$. Since x in this case is 5, that's the length of each of the square's sides. Its perimeter, then is $4 \cdot 5$, or 20 cm.

2. **14 ft.** The shortest side of a 30-60-90 triangle is half the hypotenuse. 7 ft. is half of 14 ft.

3. **18 m** The second longest side of a 30-60-90 triangle is half the hypotenuse $\sqrt{3}$. That means that in this triangle, since the side in question is $9\sqrt{3}$, 9 is half the hypotenuse. That makes the hypotenuse twice 9, or 18 m.

4. **$7\sqrt{2}$ cm** Remember, a square's diagonal cuts it into two right isosceles triangles (see question 1). When the side of a right isosceles triangle is x, the hypotenuse is $x\sqrt{2}$. That makes the diagonal $7\sqrt{2}$ cm.

5. **$36\sqrt{3}$ in²** Draw an altitude of the triangle. That makes two 30-60-90 triangles, as shown in the following figure.

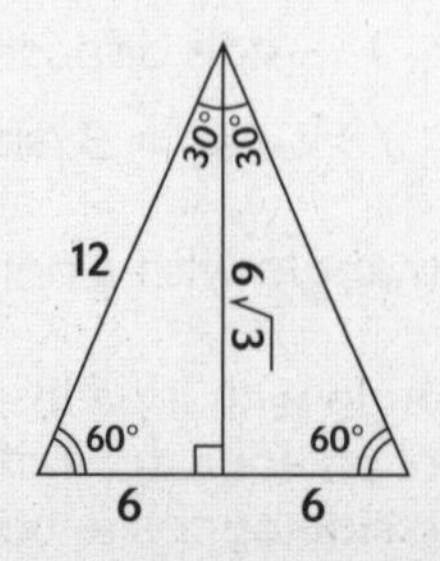

Since the side is 12, the side opposite either remaining 60° angle is half the hypotenuse $\sqrt{3}$, or $6\sqrt{3}$. Area of a triangle, you'll recall, is $\frac{1}{2}bh$.

$$A = \frac{1}{2}(12)(6\sqrt{3})$$

$$A = 6(6\sqrt{3})$$

$$A = 36\sqrt{3}\text{ in}^2$$

Chapter 8

Circles

You've probably heard that circular reasoning is not a good thing, but for understanding this chapter, nothing could be less true. You should find this chapter as easy as pumpkin π. All right, so perhaps we're being a little silly, but it was the Greek mathematicians' attempts to understand the circle that brought about the discovery of that irrational number, which we'll approximate as 3.14. Let's start out by making sure that we understand the basic terminology and notation that are used in connection with circles.

Parts of a Circle

A **circle** is a plane figure made up of all points equal in distance from a fixed point, known as the **center**. A circle is named by its center. Any line segment with an endpoint at the circle's center and its other endpoint on the circle is a **radius**. A **chord** is a line segment with endpoints lying on the circle. The longest chord in any circle is the **diameter**, which passes through the center of the circle. The length of a diameter is, for what should be obvious reasons, equal in length to twice the length of a radius, or two radii. For reasons that should be clear, in any circle, or in circles of equal size, all radii are equal in length, and all diameters are equal in length.

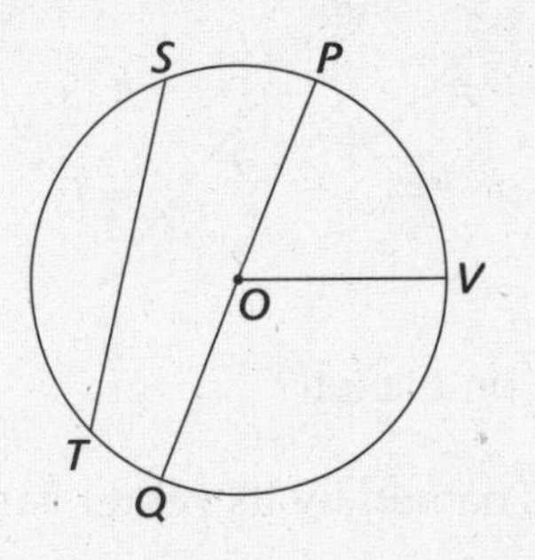

In Circle O, shown here, $\overline{ST}$ is a chord, as is $\overline{PQ}$, which is also a diameter. $\overline{OV}$ is a radius, as are $\overline{OP}$ and $\overline{OQ}$.

When a chord extends beyond the circle, it is known as a **secant**. A **tangent** is a line that touches a circle at one point only. The place where the tangent touches the circle is the **point of tangency**.

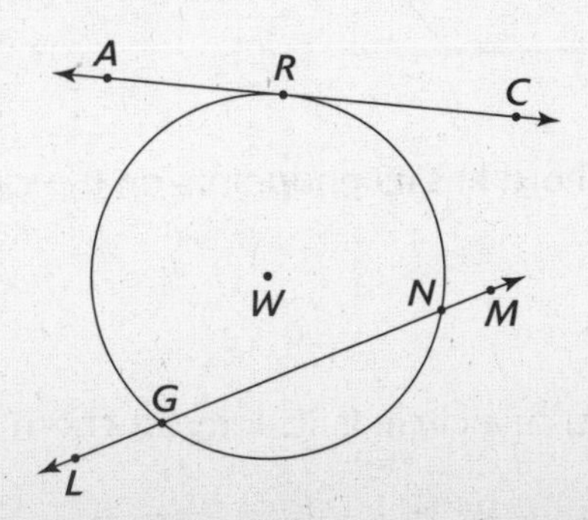

The preceding figure shows that $\overleftrightarrow{AC}$ is a tangent to Circle W, with its point of tangency at R. Chord $\overline{GN}$ is actually a part of secant $\overleftrightarrow{LM}$.

A **common tangent** is a line tangent to two circles in the same plane. If it is an **internal common tangent**, it intersects the line segment that connects the centers of the two circles (whether or not that line segment is actually drawn). An **external common tangent** does not intersect the line segment connecting the two centers. Following are examples of both types.

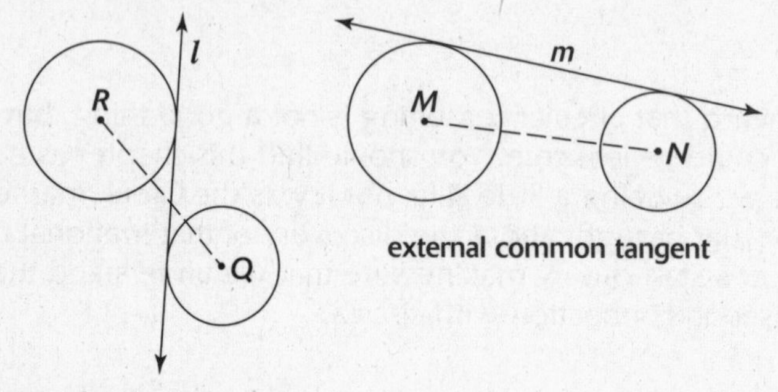

Example Problems

These problems show the answers and solutions. All problems refer to this diagram.

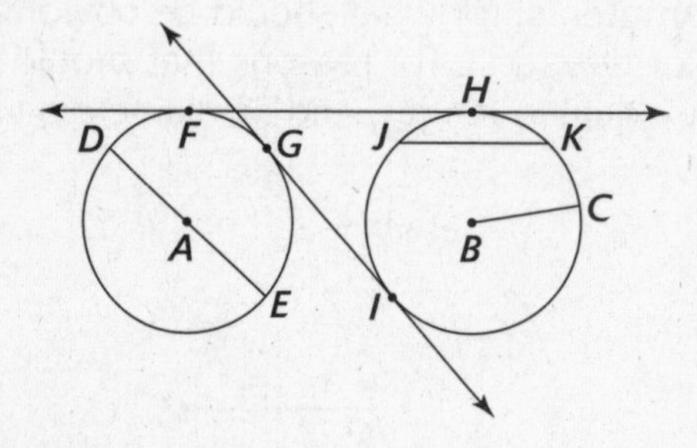

1. What is the name of the circle on the left?

   Answer:* Circle *A A circle is named by its center. In this case, the center is labeled A, so that names the circle.

2. Name an internal common tangent.

   Answer: $\overleftrightarrow{GI}$ Since $\overleftrightarrow{GI}$ touches each circle at one point, it is their common tangent. Even though no line is drawn connecting the centers of the circles, if one were drawn, $\overleftrightarrow{GI}$ would intersect it. That's what makes it an *internal* common tangent.

3. Name all chords.

   Answer: $\overline{JK}$ **and** $\overline{DE}$ A chord has its endpoints on the circle. $\overline{JK}$ and $\overline{DE}$ are the only line segments to do this.

4. Name all diameters.

   Answer: $\overline{DE}$ The diameter of a circle is its longest chord and contains the center of the circle.

5. Name an external common tangent.

Answer: $\overleftrightarrow{FH}$ Since $\overleftrightarrow{FH}$ touches each circle at one point, it is their common tangent. Even though no line is drawn connecting the centers of the circles, if one were drawn, $\overleftrightarrow{FH}$ would *not* intersect it. That's what makes it an *external* common tangent.

Work Problems

Use these problems to give yourself additional practice. All problems refer to this diagram.

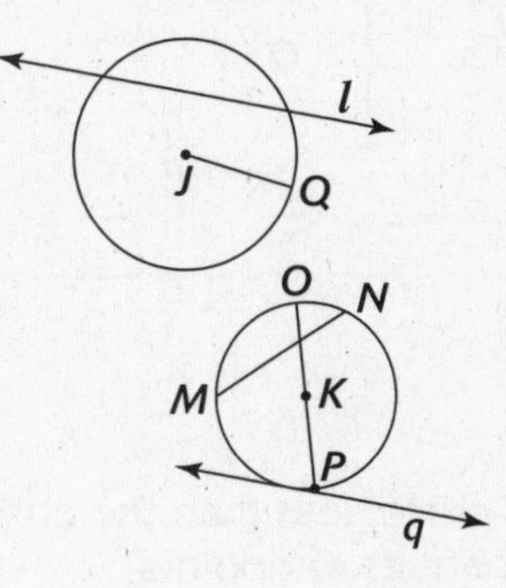

1. Name all radii in the diagram.
2. Name all secants in the diagram.
3. Name all chords in the diagram.
4. Name all tangents in the diagram.
5. Name the circle that has a tangent drawn to it.

Worked Solutions

1. $\overline{JQ}$, $\overline{KO}$, $\overline{KP}$ A radius connects the center of the circle to the circumference (the circle itself). $\overline{OP}$ is a diameter, but it contains two radii, $\overline{KO}$ and $\overline{KP}$.
2. l A secant is a line that contains a chord. Line l is the only part of the diagram that qualifies.
3. **$\overline{MN}$, $\overline{OP}$, and the part of line l within circle J** A chord is a segment with endpoints on the circle.
4. q A tangent is a line that touches the circle at a single point only.
5. **Circle *K*** Only one circle has a tangent drawn to it—the second one. It is named by its center point, *K*.

Central Angles and Arcs

Several different kinds of angle are associated with circles. Probably the one that most readily comes to mind when thinking of circles is the central angle, formed at the center by two radii.

With one of those radii held stationary, the other one can sweep, like the second hand on a clock, through a full rotation of 360°—the usual degree measure thought of as being contained by a circle.

Central Angles

As just mentioned, a central angle is formed by two radii. Its vertex is always the center of the circle.

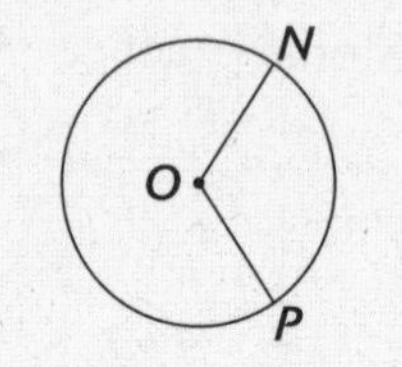

$\angle NOP$ is a central angle.

Arcs

An arc is a continuous portion of the circle, less than the entire circle. An arc is denoted by the symbol $\frown$ over the letters that designate its endpoints.

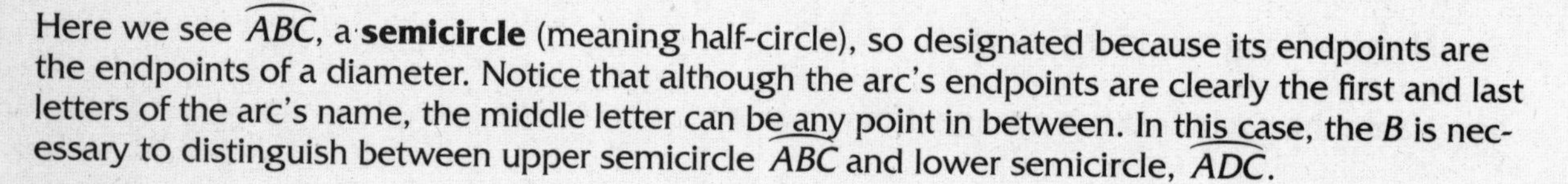

Here we see $\overset{\frown}{ABC}$, a **semicircle** (meaning half-circle), so designated because its endpoints are the endpoints of a diameter. Notice that although the arc's endpoints are clearly the first and last letters of the arc's name, the middle letter can be any point in between. In this case, the B is necessary to distinguish between upper semicircle $\overset{\frown}{ABC}$ and lower semicircle, $\overset{\frown}{ADC}$.

A **minor arc** is any arc that is less than a half-circle. Here's minor arc $\overset{\frown}{EF}$.

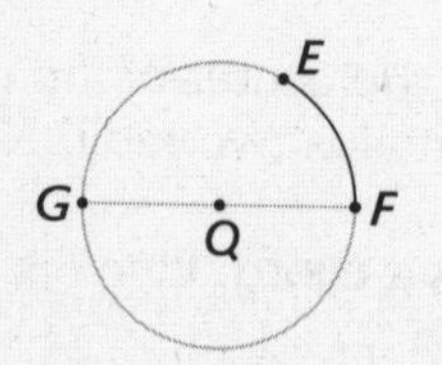

Notice that minor arcs are named with only two letters. $\overset{\frown}{EG}$ is another minor arc of circle Q.

Following, look at major arc $\overset{\frown}{PQR}$.

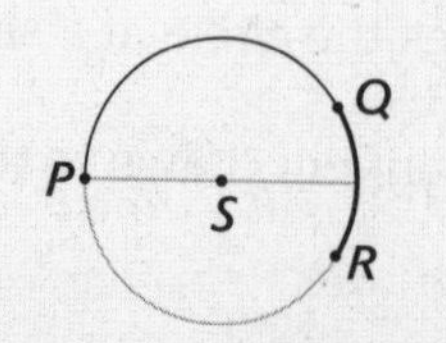

A major arc is greater than a semicircle. Like a semicircle, it also is named by three points. The arc's endpoints are the first and last letters of the arc's name, and the middle letter can be any point in between.

Arcs are measured in different ways. They may be measured by the number of degrees they contain or by their lengths. A semicircle's degree measure is 180°; its length is half the circumference of the circle. A minor arc's degree measure is defined as the same as the degree measure of the central angle that intercepts it on the circle. Its length is a portion of the circumference equal to the fraction $\frac{\angle \text{degree measure}}{360° \cdot \text{circumference}}$. The degree measure of a major arc is always 360° minus the degree measure of the minor arc that has the same endpoints. Conventionally, $\overset{\frown}{AB}$ represents the name of an arc, $m\overset{\frown}{AB}$ represents the degree measure of the arc, and $l\,\overset{\frown}{AB}$ represents the length of the arc.

Example Problems

These problems show the answers and solutions. All problems refer to the following diagram.

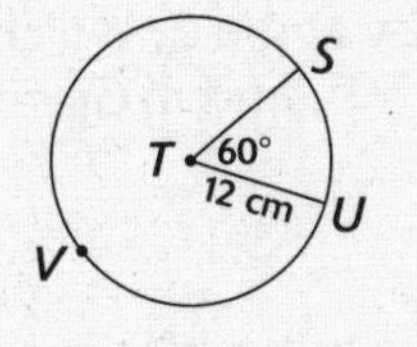

1. Find $m\overset{\frown}{SU}$.

   ***Answer:* 60°** The degree measure of an arc is the same as the degree measure of the central angle that intercepts it. That's $\angle STU$, whose degree measure is 60°.

2. Find $l\,\overset{\frown}{SU}$.

   ***Answer:* 4π cm** 60° is one-sixth of the total degree measure of the circle $\left(\frac{60°}{360°} = \frac{1}{6}\right)$. So, $l\,\overset{\frown}{SU} = \frac{1}{6}$ the circumference of circle *T*. We learned in Chapter 5 that the circumference of a circle is $2\pi r$. That's $2\pi \cdot 12$, which is 24π. $\frac{1}{6}(24\pi) = 4\pi$ cm.

3. Find $m\overset{\frown}{SVU}$ and $l\overset{\frown}{SVU}$.

   ***Answer:* 300°, 20π cm** The major arc, $\overset{\frown}{SVU}$ is both in angle measure and length what is left after minor arc $\overset{\frown}{SU}$ is subtracted from the total degree measure of the circle and from the total circumference:

   $360° - 60° = 300°$

   $24\pi - 4\pi = 20\pi$ cm

Some Common Sense Stuff

Just to keep us up to date on our formal postulates and theorems, assimilate these.

Postulate 19 (Arc Addition Postulate): If *B* is a point on $\overset{\frown}{ABC}$, then $m\overset{\frown}{AB} + m\overset{\frown}{BC} = m\overset{\frown}{ABC}$.

Theorem 66: In the same or equal circles, if two central angles have equal measures, then their corresponding minor arcs have equal measures.

Theorem 67: In the same or equal circles, if two minor arcs have equal measures, then their corresponding central angles have equal measures.

Example Problems

These problems show the answers and solutions. All problems refer to the following diagram of Circle *K*, in which $m\angle 1 = 50°$. (It is not drawn to scale.)

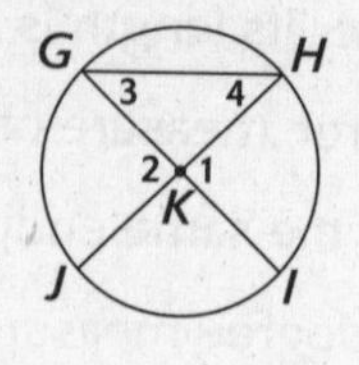

1. Find $m\overset{\frown}{HI}$.

   Answer: 50° A central angle has the same degree measure as its intercepted arc, and vice versa. $\overset{\frown}{HI}$'s central angle is $\angle 1$, so its measure is 50°.

2. Find $m\overset{\frown}{GJ}$.

   Answer: 50° $\overset{\frown}{GJ}$ is intercepted by angle 2, which is the vertical angle of $\angle 1$. Since vertical angles are equal, $m\angle 2 = 50°$, and $m\overset{\frown}{GJ} = 50°$.

3. Find $m\overset{\frown}{GH}$.

   Answer: 130° $\overline{HJ}$ is a diameter, cutting off a semicircle of 180°. $m\overset{\frown}{GJ} + m\overset{\frown}{GH} = m\overset{\frown}{HGJ}$. Remember Postulate 19, but we just found $m\overset{\frown}{GJ}$ to be 50°. So,

   $$m\overset{\frown}{GH} + 50° = 180°$$
   $$m\overset{\frown}{GH} = 180° - 50°$$
   $$m\overset{\frown}{GH} = 130°$$

4. Find $m\angle HKG$.

   Answer: 130° We just found the measure of the arc intercepted by $\angle HKG$ to be 130°. A central angle has the same degree measure as its intercepted minor arc, so $m\angle HKG = 130°$.

5. Find $m\angle 3$.

   Answer: 25° $\overline{GK}$ and $\overline{HK}$ are radii of the same circle, so they are equal in length. That makes $\triangle GHK$ isosceles, with angles opposite the equal sides equal. $\angle 3 + \angle 4 + 130° = 180°$ (the total degree measure of a triangle). $2(\angle 3) = 50°$; $\angle 3 = 25°$.

6. Find $m\angle 4$.

   Answer: 25° See Answer 5.

Arcs and Inscribed Angles

An angle that is formed by two chords with its vertex on the circle is known as an inscribed angle. It looks like this:

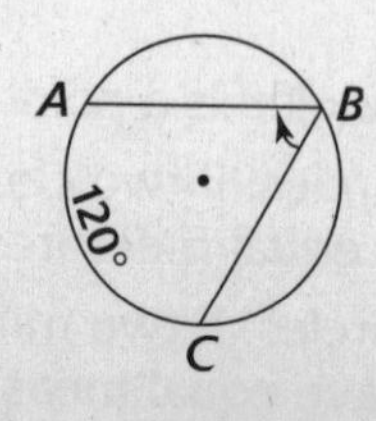

$\angle ABC$ is an inscribed angle with an intercepted arc of 120°. This is as good a place as any for a theorem.

Theorem 68: In a circle, the degree measure of an inscribed angle is half the degree measure of its intercepted arc.

That means that $m\angle ABC = \frac{1}{2} m\overset{\frown}{AC} = 60°$.

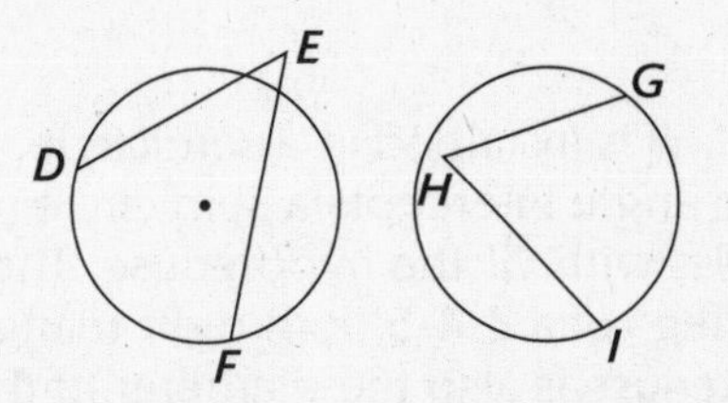

Neither $\angle DEF$ nor $\angle GHI$ are inscribed angles, since neither has its vertex on the circle.

Theorem 69: If two inscribed angles intercept the same or equal arcs, then the angles are of equal degree measure.

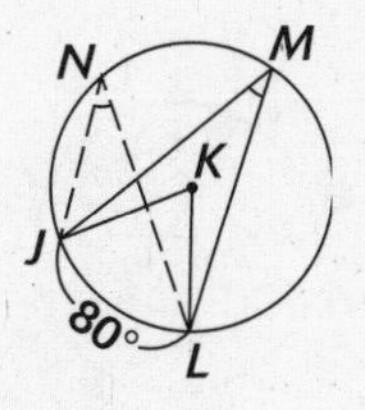

$m\angle JKL = 80°$; $m\angle JML = 40°$; $m\angle JNL = 40°$.

Theorem 70: If an inscribed angle intercepts a semicircle, then its measure is 90°.

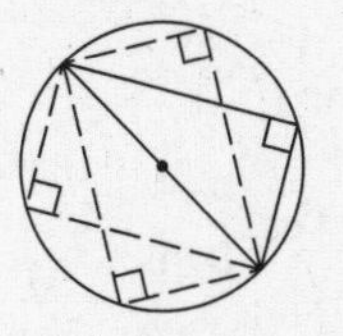

It makes sense, since a semicircle is an arc with degree measure of half 360°, or 180°.

Example Problems

These problems show the answers and solutions.

1. Central angle *AOB* in circle *O* has a degree measure of 90°. What is the measure of inscribed angle *ANB* in the same circle?

   ***Answer:* 45°** Both the central angle and the inscribed angle have the same arc, $\overset{\frown}{AB}$. Since a central angle cuts off an arc with the same degree measure as the angle, $m\overset{\frown}{AB} = 90°$. The degree measure of an inscribed angle, however, is half the intercepted arc (Theorem 68). Half of 90° is 45°.

2. Inscribed $\angle STU$ intercepts an arc of 140°. What is the degree measure of inscribed $\angle SWU$?

Answer: 70° Both inscribed angles have the same intercepted arc, $\overset{\frown}{SU}$. According to Theorem 69, if two inscribed angles intercept the same or equal arcs, then the angles are of equal degree measure. An inscribed angle, by Theorem 68, is measured by half its intercepted arc. Half of 140° is 70°.

3. $\angle GHI$ is inscribed in a semicircle. $\overline{GH}$ is 16 cm long, and $\overline{HI}$ is 12 cm long. How long is the radius of the circle?

Answer: 10 cm Since $\angle GHI$ is inscribed in a semicircle, its degree measure is 90°, by Theorem 70 (if an inscribed angle intercepts a semicircle, then its measure is 90°). That makes $\triangle GHI$ a right triangle, with $\overline{GI}$ the hypotenuse. The two given sides, 16 and 12 cm long, respectively, are the legs of a 3-4-5 (× 4) right triangle. The hypotenuse is $4 \times 5 =$ 20 cm long, but that hypotenuse is also the diameter of the circle. A radius, then, is half a diameter, or 10 cm.

Work Problems

Use these problems to give yourself additional practice.

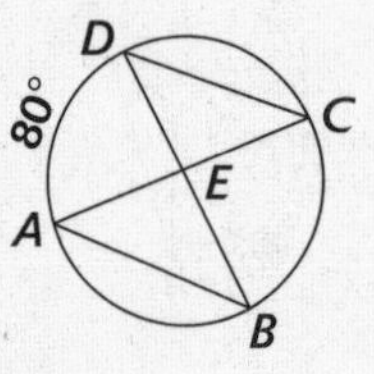

1. Find the measures of $\angle$s ACD and ABD.

2. Find the measures of $\angle$s CDB and CAB.

Problems 3–5 refer to the following diagram.

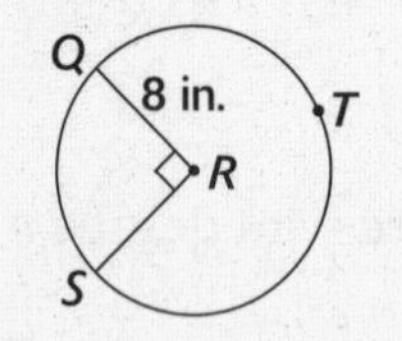

3. Find $m\overset{\frown}{QS}$.

4. Find $l\overset{\frown}{QS}$.

5. Find $m\overset{\frown}{QTS}$ and $l\overset{\frown}{QTS}$.

Worked Solutions

1. **40°, 40°** $\angle$s ACD and ABD are both intercepting $\overset{\frown}{AD}$, so according to Theorems 68 and 69, they are equal in degree measure, and that measure is half their intercepted arc. Half of 80° is 40° each.

2. **Can't tell.** $\angle AED$ intercepts $\widehat{AD}$. You might, therefore, be tempted to conclude that $m\angle AED$ is 80°. Since that's the case, $\angle CEB$, being the vertical angle of $\angle AED$, is also 80°, which makes $\widehat{CB}$ 80°. Since $\angle$s CDB and CAB are both inscribed in $\widehat{CB}$, they're 40° each. The problem with that is that **neither $\angle CEB$, nor $\angle AED$ are central angles!**

3. **90°** The degree measure of an arc is the same as the degree measure of the central angle that intercepts it. That's $\angle QRS$, whose degree measure is 90°.

4. **4π in.** 90° is one-fourth of the total degree measure of the circle $\left(\frac{90°}{360°} = \frac{1}{4}\right)$. So, $l\widehat{QS} = \frac{1}{4}$ the circumference of circle R. We learned in Chapter 5 that the circumference of a circle is $2\pi r$. That's $2\pi \cdot 8$, which is 16π. $\frac{1}{4}(16\pi) = 4\pi$ in.

5. **270°, 12π in.** The major arc, $\widehat{QTS}$ is both in angle measure and length what is left after minor $\widehat{QS}$ is subtracted from the total degree measure of the circle and from the total circumference:

   $360° - 90° = 270°$

   $16\pi - 4\pi = 12\pi$ in.

Angles Formed by Chords, Secants, and Tangents

As interesting as central angles and inscribed angles are, you're going to have to deal with chords, secants, and tangents if you really want to get to know all the angles. Let's start out with tangents and diameters.

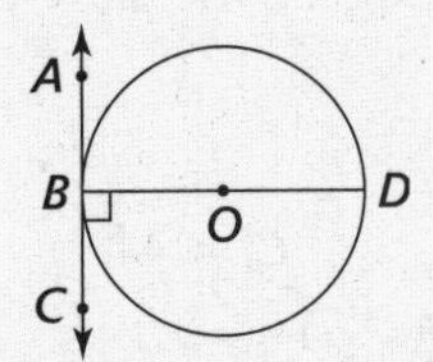

Theorem 71: If a tangent and a diameter (or radius) of a circle meet at the point of tangency, then they are perpendicular to each other.

Theorem 72: If a chord and a tangent to a circle are perpendicular at the point of tangency, then the chord is a diameter of that circle.

What these two theorems are saying, in a nutshell, is that tangents are perpendicular to diameters, and if a tangent is perpendicular to a chord, that chord must be a diameter. The preceding diagram shows it all. Interestingly, Theorem 72 is useful for finding the center of a circle.

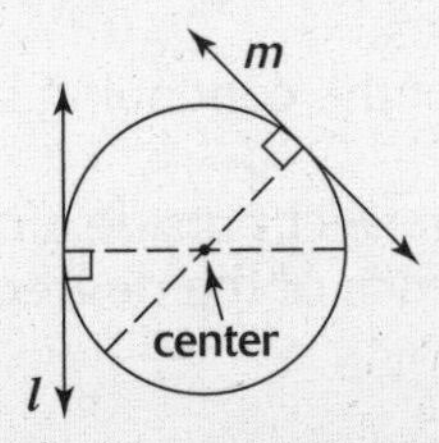

In the preceding figure, lines l and m are drawn tangent to the circle. Then chords are drawn perpendicular to each tangent at its point of tangency. Since these chords must be diameters, both must pass through the center of the circle. That's the point where the diameters intersect.

Now consider angles created by intersecting chords.

> **Theorem 73:** The degree measure of an angle formed by two chords intersecting inside a circle is equal to half the sum of the degree measures of the arcs intercepted by the angle and its vertical angle.

Referring to the preceding figure:

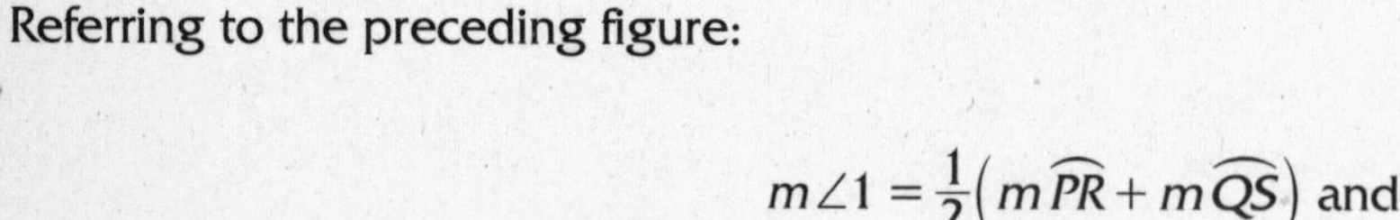

$$m\angle 1 = \frac{1}{2}\left(m\widehat{PR} + m\widehat{QS}\right) \text{ and}$$

$$m\angle 2 = \frac{1}{2}\left(m\widehat{PS} + m\widehat{QR}\right)$$

If you liked that theorem, as the saying goes, then you're sure to enjoy:

> **Theorem 74:** The degree measure of an angle formed by a tangent and a chord meeting at the point of tangency is half the degree measure of the intercepted arc.

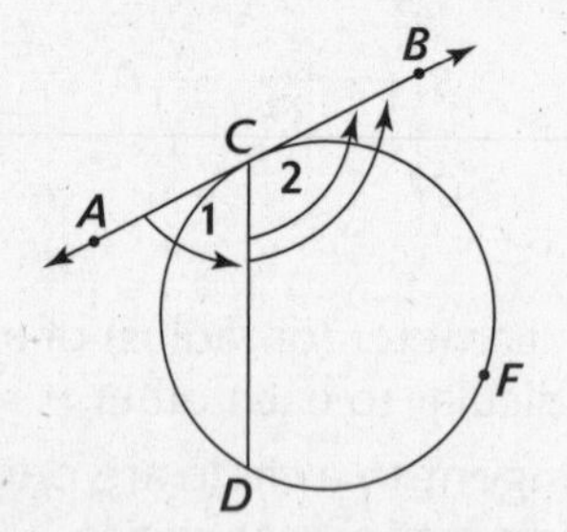

Tangent $\overleftrightarrow{AB}$ meets chord $\overline{CD}$ at C, forming $\angle$s 1 and 2 at C. According to Theorem 74:

$$m\angle 1 = \frac{1}{2} m\widehat{CD}$$
$$m\angle 2 = \frac{1}{2} m\widehat{CFD}$$

And, the final theorem for this episode of the continuing story of circles:

> **Theorem 75:** When two secants intersect outside a circle, the degree measure of the angle they form is one-half the difference of the degree measure of their intercepted arcs.

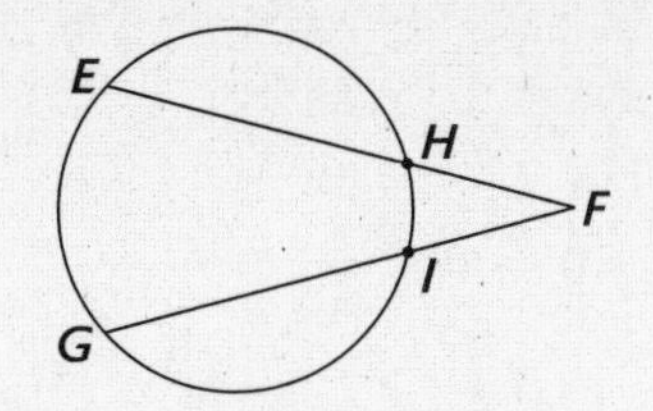

Applying Theorem 75: $m\angle EFG = \frac{1}{2}\left(m\widehat{EG} - m\widehat{HI}\right)$.

Do you feel like it's time for some practice? Then you'll be very happy about what's next.

Example Problems

These problems show the answers and solutions.

1. $m\widehat{QR} = 35°$ and $m\widehat{PS} = 55°$. Find $m\angle 1$ and $m\angle 2$.

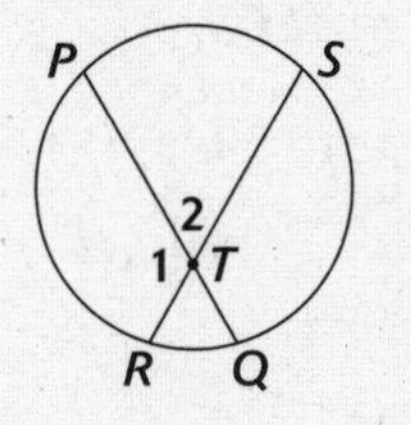

***Answer:* 135°, 45°**

By Theorem 73:

$m\angle 2 = \frac{1}{2}\left(m\widehat{PS} + m\widehat{QR}\right)$

$m\angle 2 = \frac{1}{2}(55° + 35°)$

$m\angle 2 = \frac{1}{2}(90°)$

$m\angle 2 = 45°$

Since $\angle 1$ and $\angle 2$ are supplementary:

$m\angle 1 = 180° - 45°$

$m\angle 1 = 135°$

2. $m\widehat{CD} = 140°$ and $m\widehat{CFD} = 220°$. Find $m\angle 1$ and $m\angle 2$.

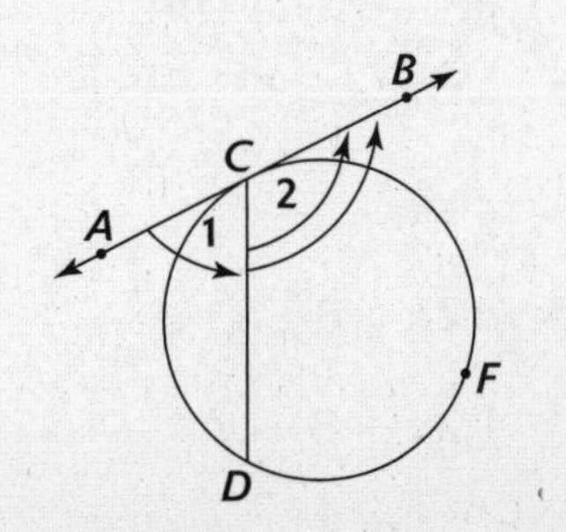

***Answer:* 70°, 110°**

By Theorem 74:

$$m\angle 1 = \frac{1}{2}m\overset{\frown}{CD}$$

$$m\angle 1 = \frac{1}{2}\cdot 140^\circ$$

$$m\angle 1 = 70^\circ$$

By Theorem 74:

$$m\angle 2 = \frac{1}{2}m\overset{\frown}{CFD}$$

$$m\angle 1 = \frac{1}{2}\cdot 220^\circ$$

$$m\angle 1 = 110^\circ$$

3. $m\overset{\frown}{EG} = 75^\circ$ and $m\overset{\frown}{HI} = 35^\circ$. Find $m\angle EFG$.

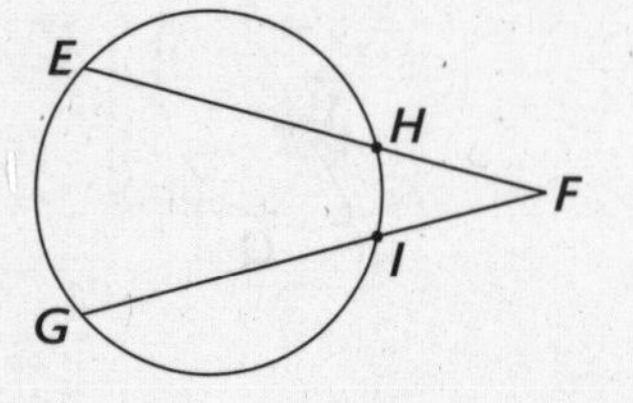

***Answer:* 20°**

By Theorem 75:

$$m\angle EFG = \frac{1}{2}\left(m\overset{\frown}{EG} - m\overset{\frown}{HI}\right)$$

$$m\angle EFG = \frac{1}{2}\left(75^\circ - 35^\circ\right)$$

$$m\angle EFG = \frac{1}{2}\left(40^\circ\right)$$

$$m\angle EFG = 20^\circ$$

4. The following figure is as marked. Find the value of x.

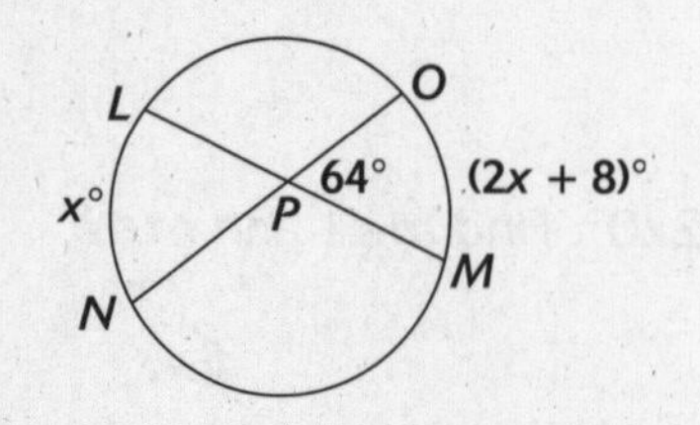

***Answer:* x = 40°**

From Theorem 73: $64 = \frac{1}{2}(2x + 8 + x)$

Multiply by 2: $128 = 1(2x + 8 + x)$

Combine like terms: $128 = 3x + 8$

Subtract 8 from both sides: $120 = 3x$

Divide by 3: $x = 40°$

Arcs and Chords

Not only angles cut off arcs on a circle. After all, a chord touches the circle at two points, and that's all it takes to make an arc. Consider these theorems:

Theorem 76: In the same or equal circles, if two chords are equal in measure, then the minor arcs with the same endpoints at those chords are equal in measure.

Theorem 77: In the same or equal circles, if two minor arcs are equal in measure, then the chords with the same endpoints as those arcs are equal in measure.

Think about those two theorems for a moment or two, and they should make sense to you. This next one takes a bit more thinking to picture, so we'll provide a diagram.

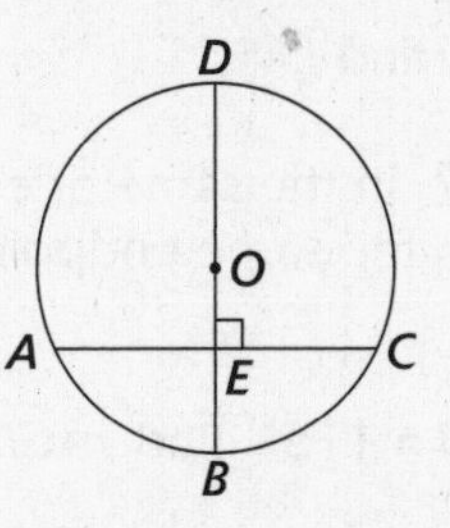

Theorem 78: If a diameter is perpendicular to a chord, then it bisects the chord and its arcs.

Theorem 78a: If a diameter bisects a chord, it is also perpendicular to that chord.

What this means is that in the preceding figure, since diameter $\overline{BD} \perp \overline{AC}$, $AE = CE$, $\widehat{AB} = \widehat{BC}$, and $\widehat{AD} = \widehat{CD}$.

The final theorems in this section are converses of each other, just as the first two were. Like the first two, if you take a moment or two to ponder them, you'll see that they make sense intuitively. You may try to reason them out if you like, but proving them is beyond the scope of this book.

Theorem 79: In the same or equal circles, if two chords are equal in measure, then they are equidistant from the center.

Theorem 80: In the same or equal circles, if two chords are equidistant from the center, they are equal in measure.

Let's see how these theorems are used.

Example Problems

These problems show the answers and solutions. Problems 1 and 2 refer to this figure:

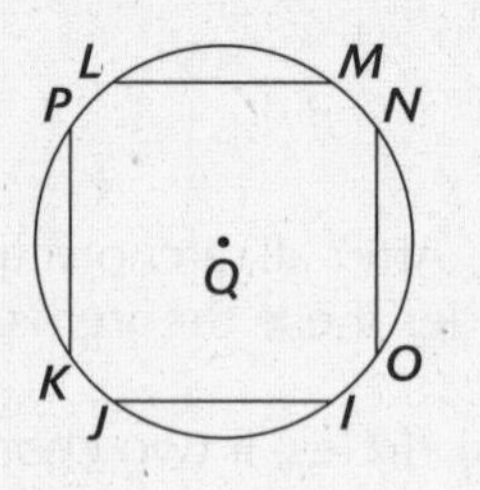

1. If $NO = IJ$ and $m\widehat{NO} = 75^\circ$, find $m\widehat{IJ}$.

   Answer: 75° According to Theorem 76: In the same or equal circles, if two chords are equal in measure, then the minor arcs with the same endpoints are equal in measure. So, $m\widehat{IJ} = m\widehat{NO} = 75^\circ$.

2. If $m\widehat{KP} = m\widehat{LM}$ and $KP = 10$ in., find LM.

   Answer: 10 in. By Theorem 77: In the same or equal circles, if two minor arcs are equal in measure, then the chords with the same endpoints are equal in measure. That means $LM = KP = 10$ in.

3. Using the following figure, $m\widehat{AB} = 115^\circ$, find $m\widehat{CD}$ and explain why.

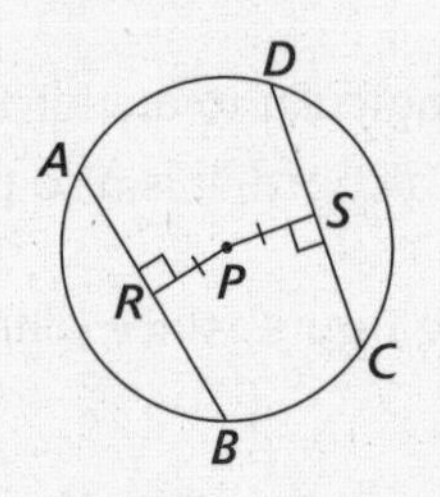

Answer: 115° because of Theorems 80 and 76.

Given: Circle P and $PR = PS$, as marked in the preceding diagram, and $m\widehat{AB} = 115^\circ$.

Prove: $m\widehat{CD} = m\widehat{AB} = 115^\circ$

Statement	*Reason*
1. $PR = PS$	1. Given
2. Chords $\overline{AB}$ and $\overline{CD}$ are equidistant from P.	2. See 1 (previous).
3. $AB = CD$	3. Theorem 80
4. $m\widehat{AB} = 115^\circ$	4. Given
5. $m\widehat{CD} = m\widehat{AB}$	5. Theorem 76
6. $m\widehat{CD} = 115^\circ$	6. Definition of equality

4. $m\widehat{JK} = 125°$; $m\widehat{LM} = 125°$; and $LM = 14$ cm. Find JK.

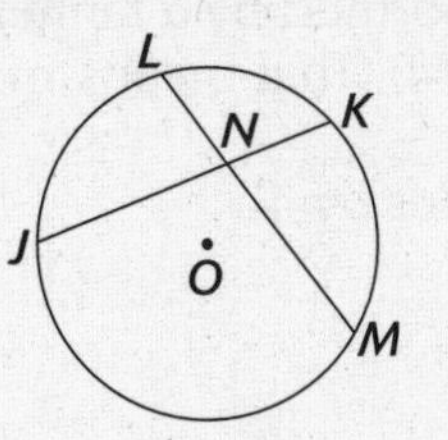

***Answer:* 14 cm** By Theorem 77: In the same or equal circles, if two minor arcs are equal in measure, then the chords with the same endpoints are equal in measure.

Work Problems

Use these problems to give yourself additional practice.

1. $m\angle 2 = 120°$. $\widehat{CFD} = 4x°$. Find x.

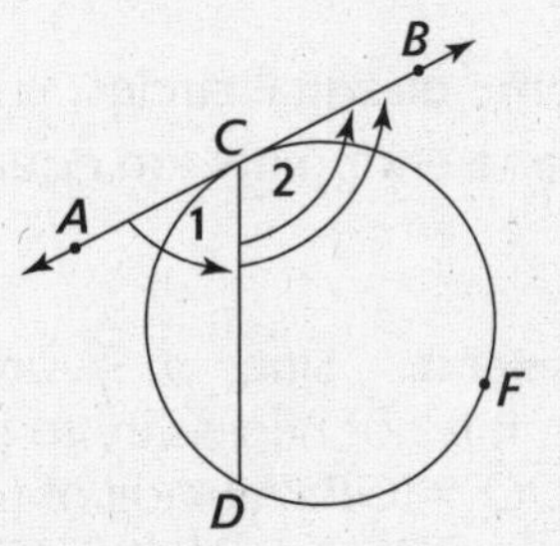

Use the following figure to solve problems 2 and 3.

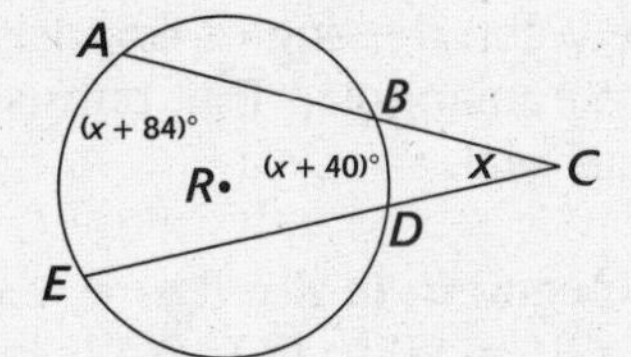

2. Find the value of x in Circle R.
3. Chords AB and ED measure 14 cm each. $m\widehat{ED} = 86°$. Find $m\widehat{AB}$.

Questions 4 and 5 pertain to the following diagram.

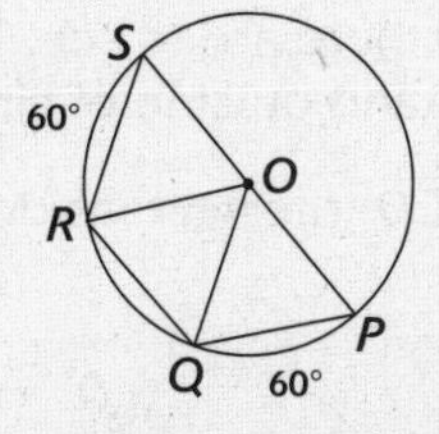

4. What can you say about $\triangle ORS$ and $\triangle OPQ$ in the figure?
5. To which other line segment in the figure is $\overline{RQ}$ equal in measure (if any)?

Worked Solutions

1. **60°** You'll recall that an angle formed by a tangent and a chord is half the angle measure of its intercepted arc (Theorem 74). That means $m\overset{\frown}{CFD} = 2 \cdot 120$.

   Since $4x = 240°$, divide both sides by 4.

   We get $x = 60°$.

2. **22°**

   From Theorem 75: $x = \frac{1}{2}[(x+84)-(x+40)]$

   Multiply by 2: $2x = (x + 84) - (x + 40)$

   Clear parentheses: $2x = x + 84 - x - 40$

   Combine like terms: $2x = 44$

   Divide by 2: $x = 22$

3. **86°** By Theorem 76: In the same or equal circles, if two chords are equal in measure, then the minor arcs with the same endpoints are equal in measure. Since $m\overset{\frown}{ED} = 86°$, $m\overset{\frown}{AB}$ must be 86°.

4. **They are congruent and equilateral.** Since $m\overset{\frown}{SR}$ and $m\overset{\frown}{PQ}$ are both 60°, $\overline{SR} = \overline{PQ}$. Although this fact is interesting, it is not necessary to prove the conclusion. Much more important is that $\overline{SO} = \overline{OR} = \overline{OQ} = \overline{OP}$. All radii of the same circle are equal. Since $\angle ROS$ and $\angle POQ$ are both 60° angles, $\triangle ROS \sim \triangle POQ$ (SAS). Both triangles are also isosceles, since their radii-sides are equal, but the measure of the angles opposite those equal sides are equal (Theorem 27). Since each triangle already has a 60° angle at O, there are 120° left in each triangle to share between the angles opposite the equal sides. That makes each angle 60°, and the triangles are equiangular. That brings us to Theorem 30: If a triangle is equiangular, then it is also equilateral.

5. **It is equal to every other line segment in the figure, except the diameter.** We already know that $\overline{SO} = \overline{OR} = \overline{OQ} = \overline{OP} = \overline{SR} = \overline{PQ}$. $\overline{RQ}$ is in the same semicircle as $\overline{SR}$ and $\overline{PQ}$. Since a semicircle describes an arc of 180°, 120° of which have already been used, $\overset{\frown}{RQ}$ must contain 60°. Then, by Theorem 77, $\overline{RQ}$ must be the same length as $\overline{SR}$ and $\overline{PQ}$, and hence, every other segment in the circle (except for $\overline{SP}$).

Segments of Chords, Secants, and Tangents

Now that we've seen all the angles that there are to see, it's time to examine some of the lines' and segments' relationships both inside and outside of circles.

In the following figure, chords $\overline{AB}$ and $\overline{CD}$ intersect inside the circle.

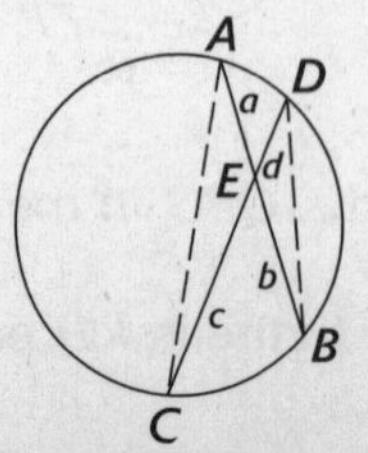

A D a E d b c B C

By adding $\overline{AC}$ and $\overline{DB}$ to the diagram, we form two triangles, which readily can be shown to be similar triangles. (Look at the intercepted arcs, and you'll see that $\angle$s *A* and *D* both intercept $\widehat{BC}$, and $\angle$s *B* and *C* both intercept $\widehat{AD}$. Since corresponding sides of similar triangles are in proportion, $\frac{a}{d} = \frac{c}{b}$. By cross-products, this yields the relationship $ab = cd$. This relationship with respect to intersecting chords is memorialized in the following theorem:

> **Theorem 81:** If two chords intersect inside a circle, then the product of the lengths of the segments of one chord equals the product of the lengths of the segments of the other chord.

In the following circle, $JK = 4$, $KI = 6$, and $HK = 3$. Let's find GK.

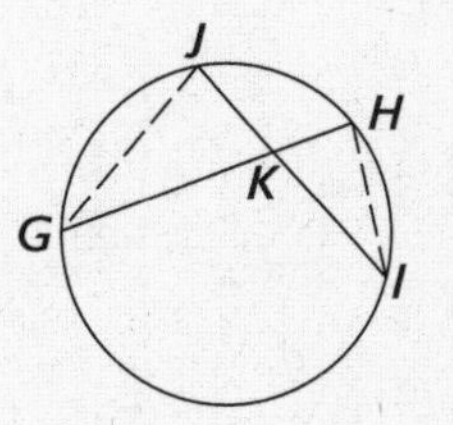

By Theorem 81: $JK \cdot KI = 3 \cdot GK$
Substituting: $4 \cdot 6 = 3 \cdot GK$
Combine terms: $3\,(GK) = 24$
Divide by 3: $GK = 8$

If you would like to check that answer, think $4 \times 6 = 24$ and $3 \times 8 = 24$. Makes sense!

Next, let's check out two secants. You can see that we have dotted-in sides to create triangles, just as we had with Theorem 81.

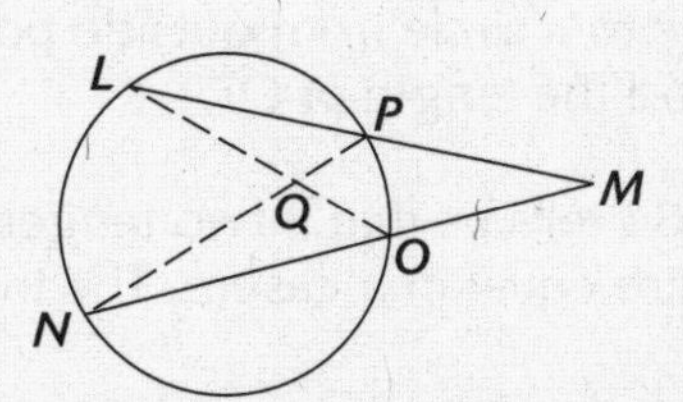

Because those triangles are easily proven to be similar, we can set up a proportion relating the sides of the triangles. By cross-products we find that $LM(MP) = MN(MO)$. Now, to be accurate, we are only showing secant segments in the preceding diagram (since technically, they continue forever), but in the interest of simplicity, we'll refer to them as secants in the following theorem.

> **Theorem 82:** If two secants to a circle intersect at an outside point, the product of one secant and its external portion is equal to the product of the other secant and its external portion.

We'll practice applying this theorem in the example problems. First, we'll look at two more potential cases. In the theorems that follow, we'll continue to refer to secants as only the portions of secants that exist between an outside point and the far side of the circle, and we'll refer to portions of tangents in the same way, which is from the outside point to the point of tangency. Although not exactly accurate, it treats the parts of those lines with which we are really concerned, while allowing theorems to be expressed in a much less cumbersome and confusing manner than might otherwise be needed.

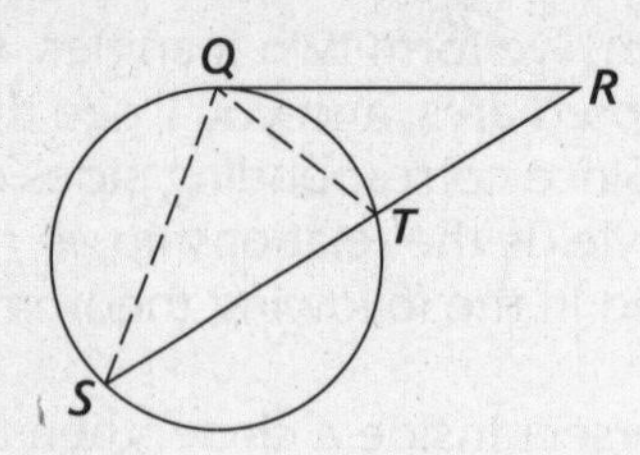

Theorem 83: If a tangent and a secant to the same circle intersect at an outside point, then the square of the measure of the tangent equals the product of the measure of the secant and its external portion.

Theorem 84: If two tangents to the same circle intersect at an outside point, then they are equal in measure.

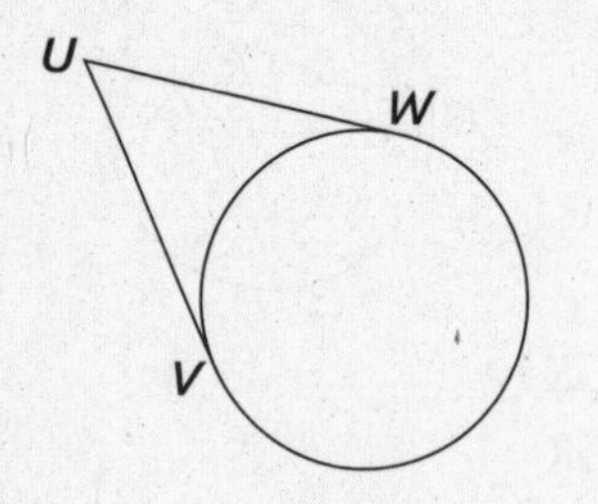

Let's try these theorems with some worked out example problems.

Example Problems

These problems show the answers and solutions.

1. $\overline{QS}$ and $\overline{QT}$ are two tangents to a circle from outside point Q. They intercept an arc of 150°. If QS is 17 cm long, find the length of QT.

   ***Answer:* 17 cm** Theorem 84 tells us that if two tangents to the same circle intersect at an outside point, then they are equal in measure. The intercepted arc has no relevance to the problem.

2. In the following figure, $LM = 36$, $MP = 8$, and $MO = 6$. Find MN.

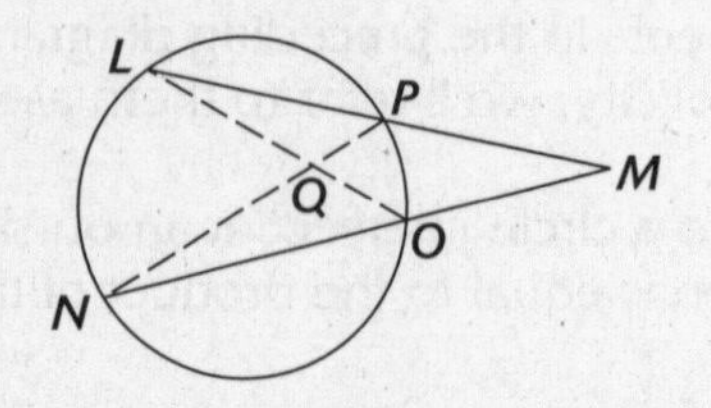

   ***Answer:* 48** By Theorem 82: If two secants to a circle intersect at an outside point, the product of one secant and its external portion is equal to the product of the other secant and its external portion.

   $MO(MN) = LM\,(MP)$

   $6(MN) = 36(8)$

   $6(MN) = 288$

   $MN = 48$

3. In the following figure, $RT = 8$ and $ST = 24$. Find QR.

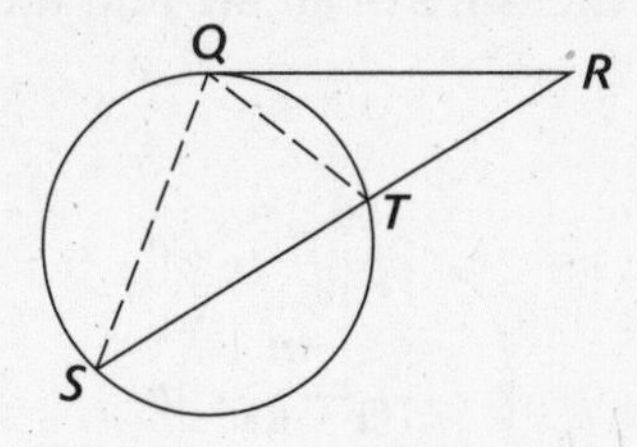

***Answer:* 16** By Theorem 83: If a tangent and a secant to the same circle intersect at an outside point, then the square of the measure of the tangent equals the product of the measure of the secant and its external portion.

Let $x = QR$.

$$x^2 = RT(ST + RT)$$

$$x^2 = 8(24 + 8)$$

$$x^2 = 8(32)$$

$$x^2 = 256$$

$$x = 16$$

4. In the following figure, $JK = 8$, $IK = 9$, and $HK = 6$. Find GK.

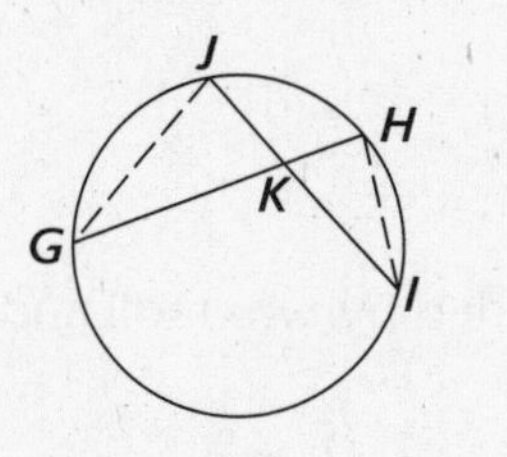

***Answer:* 12** By Theorem 81: If two chords intersect inside a circle, then the product of the segments of one chord equals the product of the segments of the other chord.

Therefore: $HK(GK) = JK(IK)$ Let $x = GK$.

Substitute: $6x = 9(8)$

Combine: $6x = 72$

Divide by 6: $x = 12$

Arc Lengths and Sectors

An **arc of a circle** is a continuous portion of the circle, less than the entire circumference in size. It is important to remember that arcs of circles have two different measurable properties. Their lengths can be measured with a flexible ruler (or they can be rolled along the floor next to a straight ruler). That was one property. The other measurable property is the degree of curvature, as a function of an arc's corresponding central angle.

Arc Length

We distinguish between the degree measure of an arc and the length of an arc by using *m* for the first and *l* for the second.

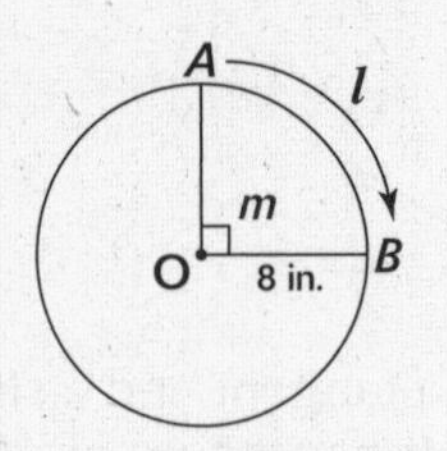

In circle *O*, $m\widehat{AB} = 90°$, since *m* central <*AOB*=90°. To determine $l\ \widehat{AB}$, we need to create a proportion. That proportion involves the relationship of the part of the circle we're seeking to find and the whole circle:

$$\frac{m \text{ central angle}}{360°} = \frac{\text{arc in question}}{\text{circumference of circle}}$$

We know the first part of the proportion we are dealing with here is $\frac{90°}{360°} = \frac{1}{4}$. We also know that the circumference of a circle is $2\pi r$, which in this case is 16π inches (*r* is 8 in.).

So, to find $l\ \widehat{AB}$: $\frac{1}{4} = \frac{l\ \widehat{AB}}{16\pi}$

Cross multiply: $4l = 16\pi$

Divide both sides by 4: $l = 4\pi$

So $l\ \widehat{AB} = 4\pi$ inches in length.

Sectors of a Circle

A **sector of a circle** is a region bounded by two radii and an arc of the circle.

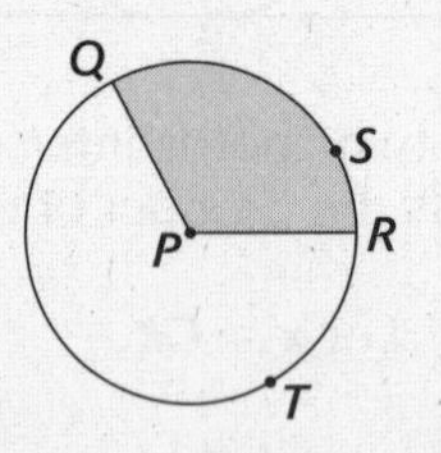

Shaded sector *QSRP* is a sector of circle *P* with $\widehat{QSR}$ and bounding radii $\overline{PQ}$ and $\overline{PR}$. *QTRP* is another sector of circle *P* with arc $\widehat{QTR}$ and bounding radii $\overline{PQ}$ and $\overline{PR}$.

The area of a sector is a part of the area of the entire circle. To find the area of a sector, the same scheme is followed as when finding the length of an arc, except this time instead of $\frac{\text{length of arc}}{\text{circumference}}$, it's $\frac{\text{area of sector}}{\text{area}}$. Suppose that the minor $\widehat{QR}$ has a degree measure of 120° and the circle has a radius of 6 cm. To find the area of *QSRP*, we write and solve the following proportion:

$$\frac{m \text{ central angle}}{360^\circ} = \frac{\text{area of sector}}{\text{area of circle}(\pi r^2)}$$

Let x = area of $QSRP$

Substitute: $\frac{120^\circ}{360^\circ} = \frac{x}{36\pi}$

Simplify: $\frac{1}{3} = \frac{x}{36\pi}$

Cross multiply: $3x = 36\pi$

Divide both sides by 3: $x = 12\pi$

So, the area of sector $QSRP = 12\pi \text{ cm}^2$.

Example Problems

These problems show the answers and solutions.

1. Prove $AB = BC$ in the following figure.

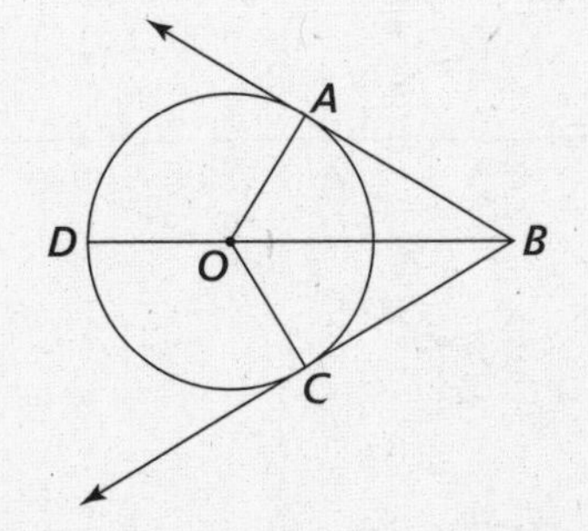

Given: Circle O with radii $\overline{AO}$, $\overline{CO}$ and tangents $\overline{AB}$, $\overline{BC}$

Prove: $AB = BC$.

Answer: **CPCTC**

Statement	***Reason***
1. $\overline{AO}$ and $\overline{CO}$ are radii to circle O.	1. Given
2. $\overline{AB}$, $\overline{BC}$ are tangents to circle O.	2. Given
3. $\overline{BA} \perp \overline{AO}$ and $\overline{BC} \perp \overline{CO}$.	3. Theorem 71
4. $\angle$s BAO and BCO are right angles.	4. Perpendiculars form right angles.
5. $AO = CO$.	5. Radii of the same circle are equal.
6. $BO = BO$.	6. Identity
7. $\triangle BAO \cong \triangle BCO$.	7. HL
8. $AB = BC$.	8. CPCTC

This is the proof of Theorem 84.

2. In circle O, which follows, $HI = 12$, $GI = 7$, and $FI = 4$. Find EI.

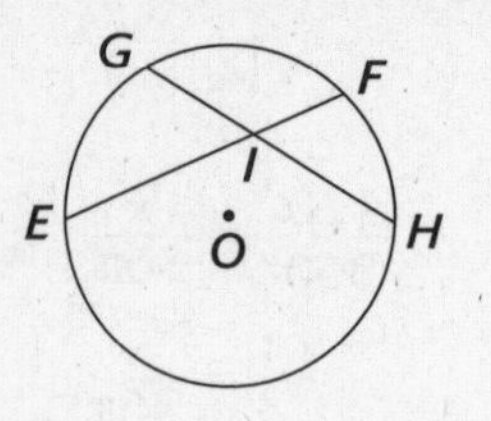

***Answer:* 21** By Theorem 81: If two chords intersect inside a circle, then the product of the segments of one chord equals the product of the segments of the other chord.

$FI \cdot EI = HI \cdot GI$ Let $x = EI$

$4x = 12 \cdot 7$

$4x = 84$

$x = 21$

3. A 20 in. long secant intersects with a tangent at a point outside the circle as shown below. 15 inches of the secant are inside the circle. How long is the tangent?

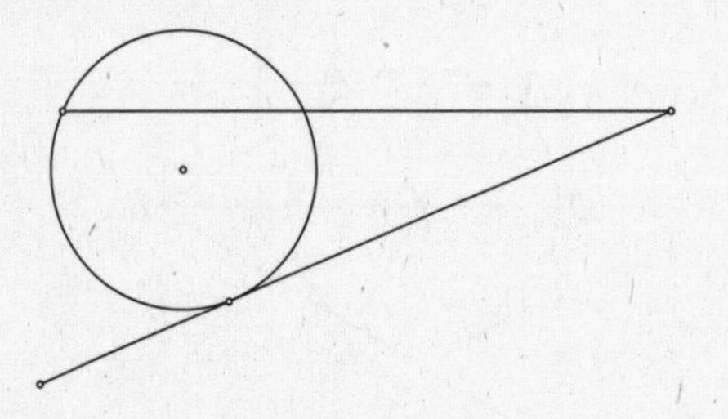

***Answer:* 10 in.** By Theorem 83: If a tangent and a secant to the same circle intersect at an outside point, then the square of the measure of the tangent equals the product of the measure of the secant and its external portion. Since the internal portion of the secant is 15 inches, the external portion must be 5 inches.

$x^2 = 5 \cdot 20$

$x^2 = 100$

$x = 10$ in.

Work Problems

Use these problems to give yourself additional practice. Use the following figure for problems 1–4.

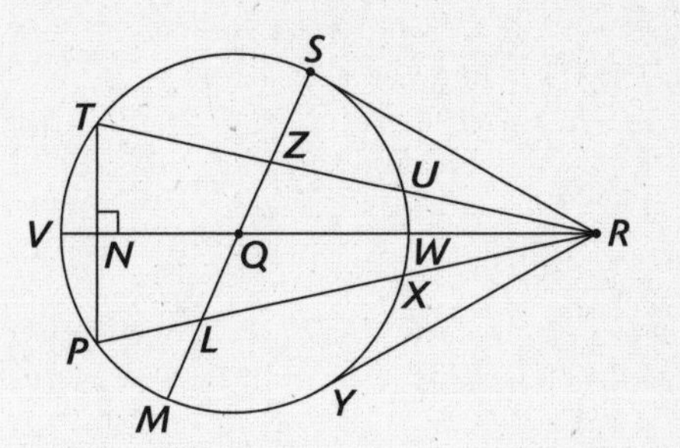

1. If $QZ = 4$, $SZ = 6$, and $RW = 16$, find SR.
2. Using all the same data as Problem 1, find RY.
3. $LM = 4$, $LP = 7$, and $LX = 8$. Find LS.
4. $RT = 32$, $RU = 9$, and $RV = 36$. Find RW.
5. In circle O, find $l\ \widehat{AXB}$ and the area of sector $AXBO$ if the radius is 9 cm.

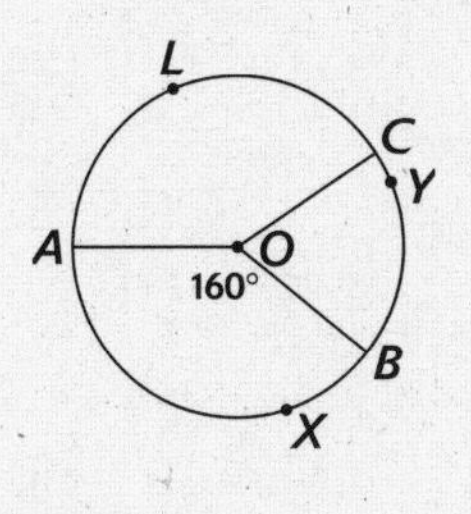

Worked Solutions

1. **24** $QZ + SZ = QS$, the length of a radius of the circle, equal to 10. $RW + QW = RQ$; but QW is a radius, and so is 10. Add it to RW to get $RQ = 26$. $\triangle QRS$ is a right triangle, since by Theorem 71, $\angle QSR$ is a right angle. Finally, either by the Pythagorean Theorem, or by Pythagorean 5, 12, 13 triple (× 2), $RS = 24$.

2. **24** By Theorem 84: If two tangents to the same circle intersect at an outside point, then they are equal in measure. RS and RY are two tangents from outside point, R.

3. **14** By Theorem 81: If two chords intersect inside a circle, then the product of the segments of one chord equals the product of the segments of the other chord. That means

$$LM \cdot LS = LP \cdot LX$$

$$4x = 7 \cdot 8$$

$$4x = 56$$

$$x = 14$$

4. **8** By Theorem 82: If two secants to a circle intersect at an outside point, the product of one secant and its external portion is equal to the product of the other secant and its external portion.

$RV \cdot RW = RT \cdot RU$

$36x = 32 \cdot 9$

$36x = 288$

$x = 8$

5. **8π cm, 36π cm²** Since the arc is equal in degree measure to the central angle that intercepts it, the arc's length is $\frac{160°}{360°}$, and its circumference is 18π ($2\pi r$).

That makes $l\,\widehat{AXB}\frac{4}{9}(18\pi) = 8\pi$ cm.

The area of the circle is $\pi r^2 = 81\pi$.

That makes the sector's area $\frac{4}{9}(81\pi) = 36\pi\text{ cm}^2$.

Chapter 9

Solid Geometry

As most intelligent people knew long before the time of Christopher Columbus, the world is *not* flat. If your house contains many containers, such as boxes, drawers, drinking glasses, or cans of food, then you probably have figured out that not all geometry is flat. In fact, your house is an example of two or more nonflat geometric figures. Those geometric figures that have a third dimension to them, known alternatively as height or depth, are called solids. Even though a box may be hollow, for our purposes it is referred to as a solid. The main difference between plane figures and solid figures is that, unlike plane figures, solids occupy space. We need to be able to measure both the surface areas of those figures (to determine how much paint will be needed to cover them) and their volume—the space that they take up or the amount that they can hold for storage.

Prisms

Prisms are geometric solids that come in many different shapes and, of course, an infinite number of sizes. Probably most familiar to you is the rectangular prism—essentially the shape of almost every rectangular carton that you've ever seen. All prisms share certain characteristics, which are listed in this section.

All prisms have two **bases**, which are polygons congruent to each other and in parallel planes. The shape of the base gives the prism its name. That is, a prism with triangular bases is a triangular prism; one with rectangles as its bases is a rectangular prism; and so on. The **lateral edges** are the edges of the prism formed by the line segments that connect the vertices of the bases.

Oblique vs. Right Prisms

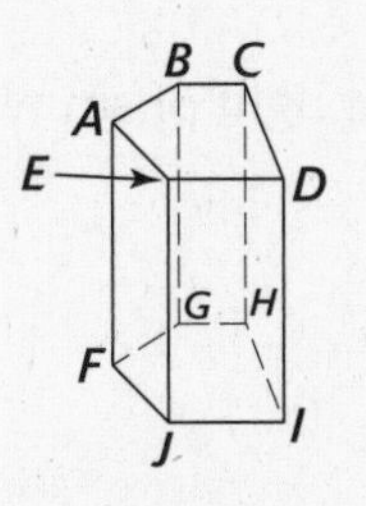

$\overline{AF}$, $\overline{EJ}$, $\overline{DI}$, $\overline{CH}$, and $\overline{BG}$ are the lateral edges of the pentagonal prism in the preceding figure. The parallelograms formed by the lateral edges are known as the **lateral faces**. *DEJI* is one lateral face of the prism; *AEJF* is another one. An **altitude** of a prism is a segment perpendicular to the planes of the bases and having an endpoint in each base. A prism whose lateral edges are perpendicular to the bases is called a **right prism** (since the edges form right angles). The prism in the preceding figure is technically a right pentagonal prism. Prisms *a* and *b* in the following figure are right rectangular and right triangular prisms, respectively. Prism *c* has lateral edges that are *not* perpendicular to the bases. It's known as an **oblique prism**. On the oblique rectangular prism, *c*, the altitude is the distance marked *h*.

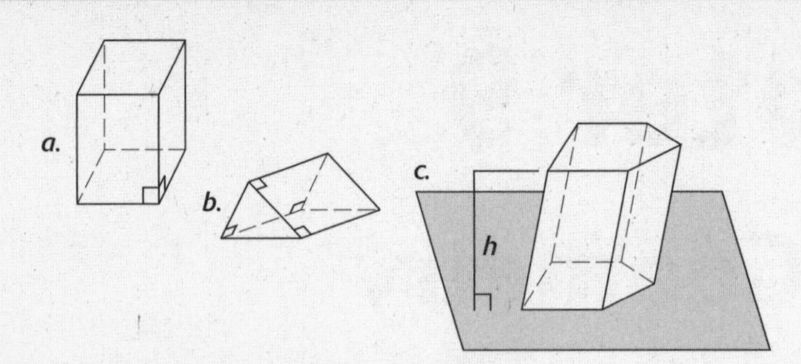

Right prisms, as already noted, are three-dimensional figures in which the lateral edges form right angles with the bases. In a right prism, any lateral edge is also an altitude. Two types of surface area are relevant to prisms. We'll look at one of them at a time.

Each lateral face of a right prism is a rectangle, whose area may be found by multiplying its height by the length of the edge of the base that is perpendicular to the height. The lateral area of the right prism, therefore, is the sum of the areas of all of its lateral faces.

> **Theorem 85:** The lateral area, *LA*, of a right prism of altitude *h* and perimeter *p* is given by the equation, $LA_{\text{prism}} = ph$ units^2.

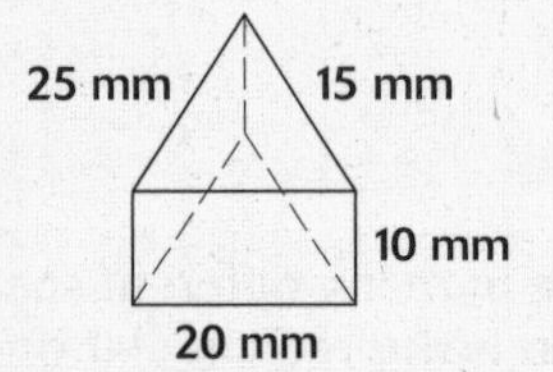

To find the perimeter of this right triangular prism, add 15 mm + 20 mm + 25 mm.

$p = 60$ mm

To find lateral area: $LA = ph$

By substituting: $LA = 60 \cdot 10$

Multiply: $LA = 600 \text{ mm}^2$

The total area of a right prism is the lateral area plus the sum of the areas of the two bases. Of course, since the bases are congruent, their areas are equal.

> **Theorem 86:** The total area, *TA*, of a right prism with lateral area *LA* and a base area *B* is given by the equation,
>
> $TA_{\text{prism}} = LA + 2B$ or
>
> $TA_{\text{prism}} = ph + 2B$

Example Problems

These problems show the answers and solutions.

1. Find the total area of the right rectangular prism shown here.

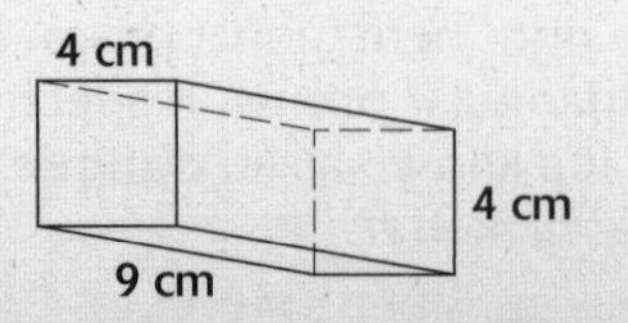

***Answer:* 176 cm^2**

$B = 16, h = 9, p = 16$

To find total area: $TA = ph + 2B$

Substitute: $TA = 16(9) + 2(16)$

Multiply: $TA = 144 + 32$

Add: $TA = 176 \text{ cm}^2$

Notice that the height has nothing to do with the direction in which the figure is oriented, although that's clearer in the case of the next figure.

2. Find the lateral area of the right hexagonal prism shown in the following figure. Each side of the base is 5 inches, and its height is 12 inches.

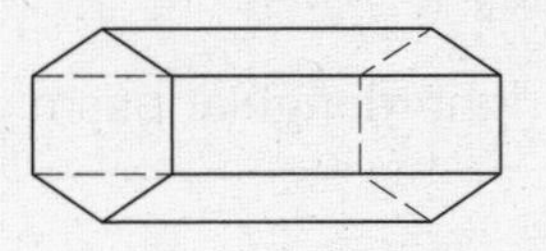

***Answer:* 360 in^2**

$$s = 5, h = 12, p = 6 \cdot 5 = 30$$

To find lateral area: $LA = ph$

By substituting: $LA = 30 \cdot 12$

Multiply: $LA = 360 \text{ in}^2$

Notice the height is horizontal!

3. Find the total area of a right triangular prism with legs 3 mm and 4 mm and a height of 20 mm. The base is a right triangle.

***Answer:* 252 mm^2**

Since the base is a right triangle, either of the two legs can be used as an altitude and the other as the base to find B.

$B = \frac{1}{2}bh$
$B = \frac{1}{2}(3)(4)$
$B = \frac{1}{2}(12)$
$B = 6$

Next, since it's a right triangle with legs 3 and 4, the hypotenuse must be 5 (Pythagoras again).

That means: $p = 3 + 4 + 5$

$p = 12$ It's time for the formula.

To find total area: $TA = ph + 2B$

Substitute: $TA = 12(20) + 2(6)$

Multiply: $TA = 240 + 12$

Add: $TA = 252 \text{ mm}^2$

Work Problems

Use these problems to give yourself additional practice.

1. Find the total area of a cube that is 6 inches wide.

2. Find the total surface area of a right triangular prism with sides of 5 ft., 12 ft., and 13 ft. and a height of 14 ft.

3. Find the lateral surface area of a rectangular prism that is 5 inches long, 7 inches wide, and 6 inches high.

4. Find the lateral surface area of a right pentagonal prism with sides of 5, 6, 7, 6, and 5 cm and a height of 8 cm.

5. Find the total surface area of a right isosceles trapezoidal prism with the trapezoid having bases of 7 m and 12 m, a leg of 6 m, and a height of 4 m. The prism's altitude is 9 m.

Worked Solutions

1. **216 in^2** We could use $TA = ph + 2B$, but why bother? A cube has 6 identical faces, each with an area of s^2. Since we are told the cube is "6 inches wide," one face has an area of 36. For the total area, multiply that by 6 and add the units: $6 \cdot 36 = 216 \text{ in}^2$.

2. **480 ft^2**

Each base has an area of

$$B = \frac{1}{2}bh$$

$$B = \frac{1}{2}(5)(12)$$

$$B = \frac{1}{2}(60)$$

$$B = 30$$

You did notice it was a right triangle?!

Perimeter: $p = 5 + 12 + 13 = 30$

To find the total area: $TA = ph + 2B$

Substitute: $TA = 30(14) + 2(30)$

Multiply: $TA = 420 + 60$

Add: $TA = 480 \text{ ft}^2$

3. **144 in²**

Find perimeter: $p = 2l + 2w$

$p = 2(5) + 2(7)$

$p = 10 + 14$

$p = 24, h = 6,$

To find lateral area: $LA = ph$

By substituting: $LA = 24 \cdot 6$

Multiply: $LA = 144 \text{ in}^2$

4. **232 cm²**

Find perimeter: $p = s_1 + s_2 + s_3 + s_4 + s_5$

$p = 2(5) + 2(6) + 7$

$p = 10 + 12 + 7$

$p = 29, h = 8$

To find the lateral area: $LA = ph$

By substituting: $LA = 29 \cdot 8$

Multiply: $LA = 232 \text{ cm}^2$

5. **355 m²** Each base has an area of $B = \frac{1}{2}(b_1 + b_2)h$ (Remember, it's a trapezoid.)

$B = \frac{1}{2}(12 + 7)(4)$

$B = \frac{1}{2}(76)$

$B = 38$

Perimeter: $p = 7 + 12 + 6 + 6 = 31$ (Each leg is 6 m.)

To find the total area: $TA = \text{ph} + 2B$

Substitute: $TA = 31(9) + 2(38)$

Multiply: $TA = 279 + 76$

Add: $TA = 355 \text{ m}^2$

Volume of a Prism

Although both lateral and total area are measurements of the surface of a solid, the interior space of a solid is what distinguishes it from a plane figure. That interior space is known as **volume** and can be measured. The unit of measurement for volume is the **unit cube**. A **cube**, you'll recall, is a square right prism with all of its edges of equal length. A unit cube is a cube whose edges are all of length 1. It may be 1 mm, 1 cm, 1 foot, and so on. The volume of any solid is determined by the number of unit cubes it takes to fill it completely. Volume is expressed in cubic units, or units^3.

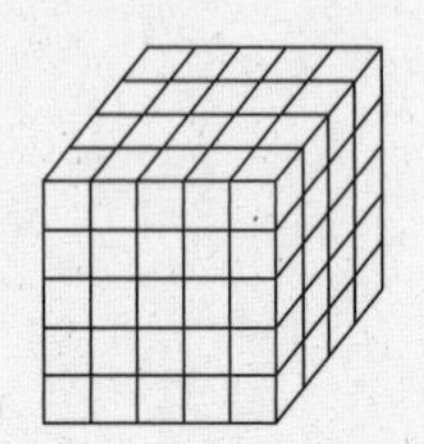

Theorem 87: The volume of a right prism with a base area B, and an altitude h, is given by the equation

$$V_{\text{prism}} = Bh \text{ units}^3$$

The volume of the right rectangular prism shown here is equal to the number of 1 cm cubes that fill it. It is 5 cm wide, 5 cm high, and 4 cubes deep. Its volume may be found by the formula given or by multiplying those dimensions: $5 \times 5 \times 4 = 100$ cm^3.

Example Problems

These problems show the answers and solutions.

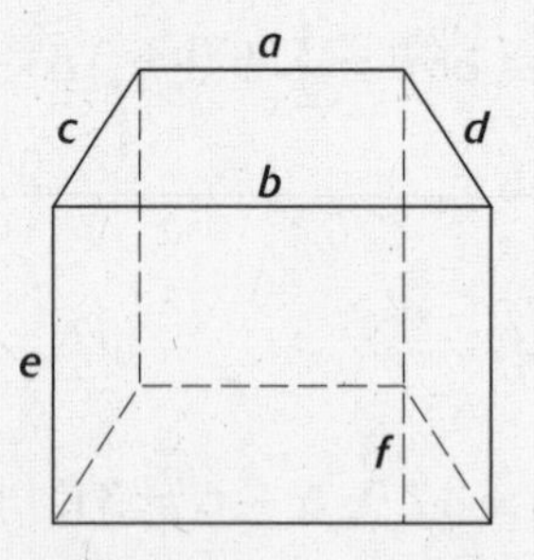

1. $a = 12$ in., $b = 20$ in., $c = 7$ in., $d = 7$ in., $e = 14$ in., and $f = 6$ in. Find the volume of the right trapezoidal prism.

Answer: 1344 in^3

Each base has an area of

$$B = \frac{1}{2}(b_1 + b_2)h$$

$$B = \frac{1}{2}(12 + 20)(6)$$

$$B = \frac{1}{2}(192)$$

$$B = 96$$

Next, ready the volume formula: $V = Bh$ units^3

Substitute: $V = 96 \cdot 14$

Multiply: $V = 1344 \text{ in}^3$

Notice that dimensions $c = 7$ in. and $d = 7$ in. had nothing to do with the problem.

2. A 10 ft. high right rectangular prism has a base that is 3 ft. by 4 ft. What is the prism's volume?

***Answer:* 120 ft³**

Each base has an area of: $B = lw$

$B = 3 \cdot 4$

$B = 12$

Next, we set up the volume formula: $V = Bh \text{ units}^3$

Substitute: $V = 12 \cdot 10$

Multiply: $V = 120 \text{ ft}^3$

3. A 6 yard high right triangular prism has a base that is a right triangle, 6 yards by 8 yards by 10 yards. What is the prism's volume?

***Answer:* 144 yd³**

Each base has an area of

$B = \frac{1}{2}bh$

$B = \frac{1}{2}(8)(6)$

$B = \frac{1}{2}(48)$

$B = 24$

Next, ready the volume formula: $V = Bh \text{ units}^3$

Substitute: $V = 24 \cdot 6$

Multiply: $V = 144 \text{ yd}^3$

Right Circular Cylinders

A prism that has circular bases is known as a cylinder. When the segments joining the bases are perpendicular to them, it is a **right circular cylinder**.

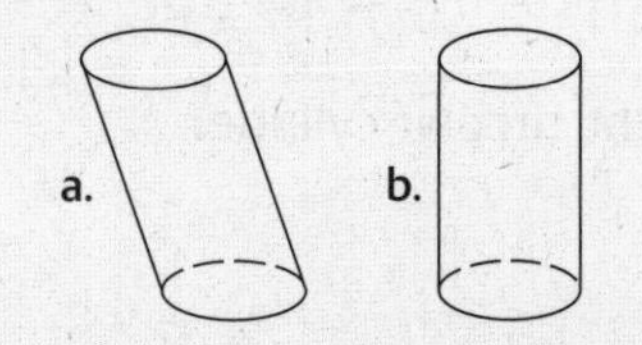

Both *a* and b in the preceding figure are cylinders. Figure *a* is not a right circular cylinder, but Figure *b* is. We find the lateral area, total area, and volume for a right circular cylinder in the same way that we do for right prisms. Picture a cylindrical can of peas with both lids removed and with the tubular part rolled out flat to form a rectangle, as in the following figure.

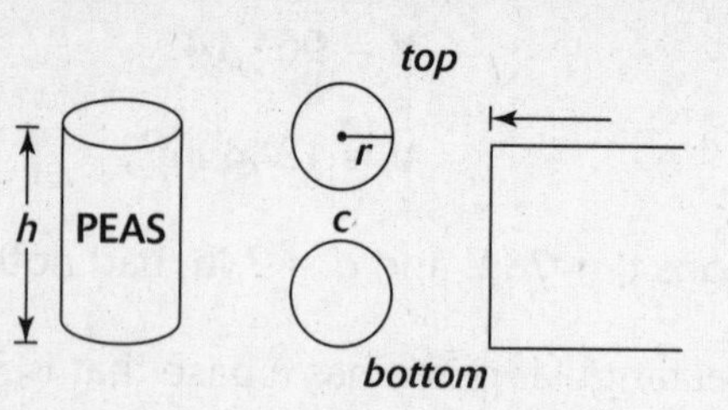

The lateral area's width is the height of the can, and its length is the circumference of the can. The bases' areas, of course, are found using the formula for the areas of circles. All of this gives rise to the following three theorems:

Theorem 88: The lateral area, LA, of a right circular cylinder with a base circumference of C and altitude h is given by the equation,

$$LA_{\text{cylinder}} = Ch \text{ units}^2 \text{ or}$$
$$LA_{\text{cylinder}} = 2\pi rh \text{ units}^2$$

Theorem 89: The total area, TA, of a right circular cylinder with lateral area, LA, and a base area B is given by the equation,

$TA_{\text{cylinder}} = LA + 2B \text{ units}^2$ or
$TA_{\text{cylinder}} = 2\pi rh + 2\pi r^2 \text{ units}^2$ or
$TA_{\text{cylinder}} = 2\pi r(h + r) \text{ units}^2$

Theorem 90: The volume of a right circular cylinder, V, with base area B and altitude h is given by the equation,

$V_{\text{cylinder}} = Bh \text{ units}^3$ or
$V_{\text{cylinder}} = \pi r^2 h \text{ units}^3$

Example Problems

These problems show the answers and solutions. All problems refer to the following figure, in which $h = 8$ cm and $r = 3$ cm.

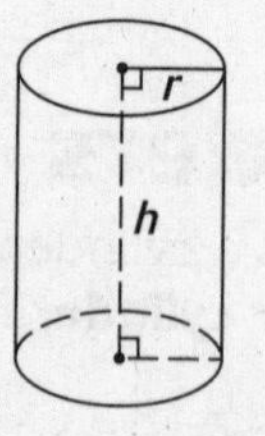

1. Find the lateral area of the right circular cylinder.

***Answer:* 48π cm²**

LA formula: $LA = 2\pi rh \text{ units}^2$

Substitute: $LA = 2\pi(3)(8)$

Multiply once: $LA = 2\pi(24)$

. . . and again: $LA = 48\pi$ cm^2

2. Find the total area of the right circular cylinder.

***Answer:* 66π cm²**

First find B: $B = \pi r^2$ units^2

Substitute: $B = \pi(3)^2$

Multiply: $B = 9\pi$ cm^2

TA formula: $TA = LA + 2B$ units^2

Substitute: $TA = 48\pi + 2(9\pi)$

Multiply: $TA = 48\pi + 18\pi$

Add: $TA = 66\pi$ cm^2

3. Find the volume of the right circular cylinder.

***Answer:* 72π cm³**

V formula: $V = Bh$ units^3

Substitute (from previous): $V = (9\pi)(8)$

Multiply: $V = 72\pi$ cm^3

Work Problems

Use these problems to give yourself additional practice. Use the following figure and data to solve problems 1 and 2.

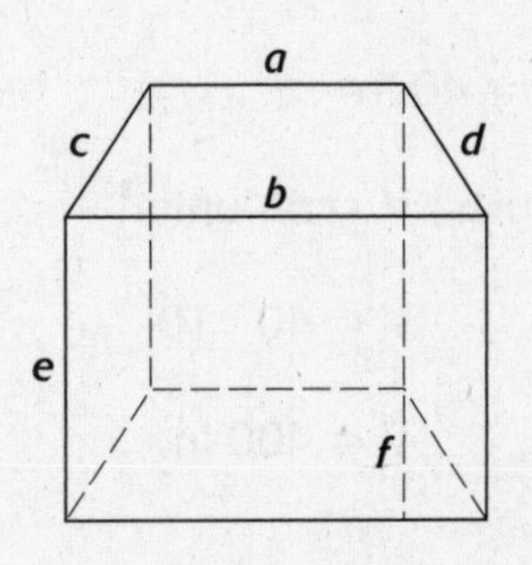

1. $a = 8$ in., $b = 12$ in., $c = 5$ in., $d = 5$ in., $e = 10$ in., and $f = 4$ in. Find the total area of the right trapezoidal prism.

2. Find the volume of the right trapezoidal prism.

Use the following diagram, with $h = 15$ cm and $r = 6$ cm, for problems 3 through 5.

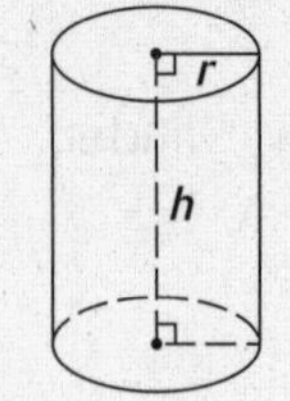

3. Find the lateral area of the right circular cylinder.
4. Find the total area of the right circular cylinder.
5. Find the volume of the right circular cylinder.

Worked Solutions

1. **380 in²**

Each base has an area of

$$B = \frac{1}{2}(b_1 + b_2)h$$
$$B = \frac{1}{2}(8 + 12)(4)$$
$$B = \frac{1}{2}(80)$$
$$B = 40 \text{ in}^2$$

Next, write the total area formula: $TA = ph + 2B$

$$p = 8 + 12 + 5 + 5 = 30 \text{ in.}$$

Substitute: $TA = 30(10) + 2(40)$

Multiply: $TA = 300 + 80$

Add: $TA = 380 \text{ in}^2$

2. **400 in³**

We have already found that $B = 40 \text{ in}^2$

Next, set up the volume formula: $V = Bh \text{ units}^3$

Substitute: $V = 40 \cdot 10$

Multiply: $V = 400 \text{ in}^3$

3. **180π cm²**

LA formula: $LA = 2\pi rh \text{ units}^2$

Substitute: $LA = 2\pi(6)(15)$

Multiply once: $LA = 2\pi(90)$

. . . and again: $LA = 180\pi \text{ cm}^2$

4. **252π cm^2**

First find B: $B = \pi r^2$ units^2

Substitute: $B = \pi(6)^2$

Multiply: $B = 36\pi$ cm^2

TA formula: $TA = LA + 2B$ units^2

Substitute: $TA = 180\pi + 2(36\pi)$

Multiply: $TA = 180\pi + 72\pi$

Add: $TA = 252\pi$ cm^2

5. **540π cm^3**

Write the V formula: $V = Bh$ units^3

Substitute (from previous): $V = (36\pi)(15)$

Multiply: $V = 540\pi$ cm^3

Pyramids

First, the terminology: A **pyramid** is a solid that has a **base** in the shape of a polygon, each vertex of which is joined to a single point in a plane other than that of the base. This point is known as the **vertex** of the pyramid. The pyramid's sides, known as its **lateral faces**, are all triangular in shape and meet at the vertex. The segments where the lateral faces meet are called **lateral edges**. The segment that runs from the vertex and intersects the base at a right angle is the pyramid's **altitude**.

A pyramid whose base is a regular polygon and whose lateral edges are all equal in length is known as a regular pyramid. Some are pictured here.

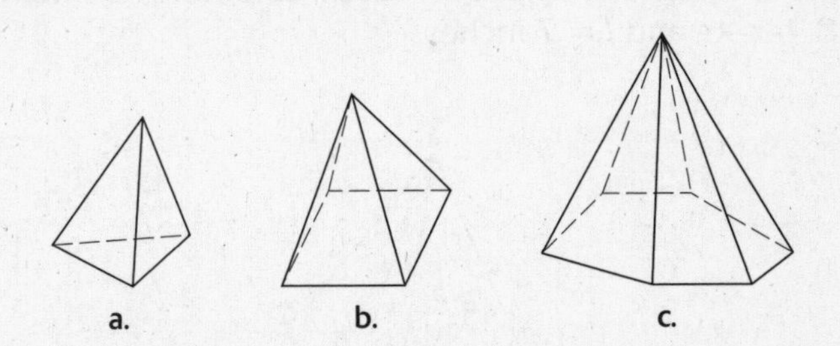

On the left, *a*, is a regular triangular pyramid; in the center, *b*, is a regular square pyramid; and on the right, c, is a regular hexagonal pyramid. Note that a pyramid is named by the shape of its base.

In a regular pyramid, the sides are all congruent isosceles triangles. The altitude of any of those triangles is the **slant height** of the pyramid, not to be confused with the altitude of the pyramid.

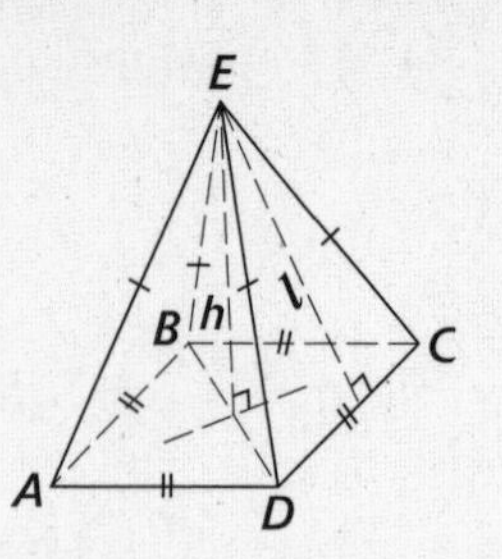

In the preceding figure, h is the altitude of the pyramid, l is the slant height, and E is the vertex of the pyramid. Square $ABCD$ is the base; $\triangle ABE$, $\triangle BCE$, $\triangle CDE$, and $\triangle ADE$ are lateral faces; and $\overline{AE}$, $\overline{BE}$, $\overline{CE}$, and $\overline{DE}$ are lateral edges. Just like the other solids we've looked at, pyramids have lateral area, total area, and volume. Fortunately, theorems are available to cover finding them:

> **Theorem 91:** The lateral area, LA, of a regular pyramid with slant height l and base perimeter p is given by the equation,
>
> $$LA_{\text{pyramid}} = \frac{1}{2}pl \text{ units}^2$$
>
> **Theorem 92:** The total area, TA, of a regular pyramid with lateral area LA and a base area B is given by the equation,
>
> $$TA_{\text{pyramid}} = LA + B \text{ units}^2 \text{ or}$$
> $$TA_{\text{pyramid}} = \frac{1}{2}pl + B \text{ units}^2$$
>
> **Theorem 93:** The volume of a regular pyramid with a base area B, and an altitude h, is given by the equation:
>
> $$V_{\text{pyramid}} = \frac{1}{3}Bh \text{ units}^3$$

Example Problems

These problems show the answers and solutions. All three problems refer to the following diagram. $AD = 14$, $DE = 8$, $h = 6$, and $l = 7$ inches.

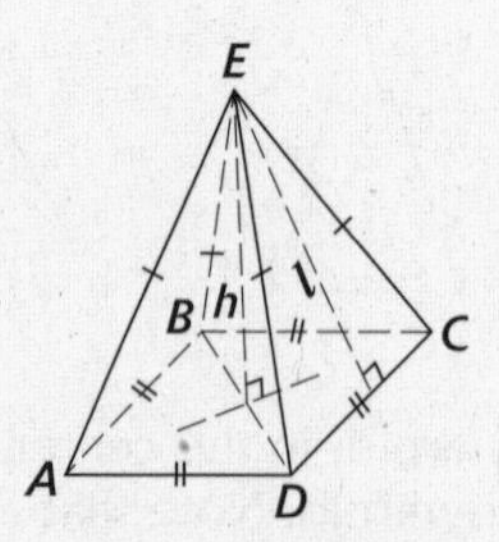

1. Find the lateral area of the regular square pyramid.

***Answer:* 196 in^2**

The perimeter of the base, p, is found by multiplying $4 \times 14 = 56$.

Write the LA formula: $LA = \frac{1}{2}pl$ units^2

Substitute: $LA = \frac{1}{2}(56)(7)$

Multiply by $\frac{1}{2}$: $LA = 28(7)$

Multiply: $LA = 196$ in^2

2. Find the total area of the regular square pyramid.

***Answer:* 392 in^2**

First find B: $B = (AD)^2$ units^2

Substitute: $B = (14)^2$

Multiply: $B = 196$ in.^{2}

Use the TA formula: $TA = LA + B$ units^2

Substitute: $TA = 196 + 196$

Add: $TA = 392$ in^2

3. Find the volume of the regular square pyramid.

***Answer:* 392 in^3**

Use the V formula: $V = \frac{1}{3}Bh$ units^3

Substitute (from previous): $V = \frac{1}{3}(196)(6)$

Multiply 6 by $\frac{1}{3}$: $V = (196)(2)$

Multiply: $V = 392$ in^3

Right Circular Cones

A right circular cone is a dead ringer for a regular pyramid, but the base is a circle instead of a polygon. Look at the following diagram, and you'll see how similar the two are.

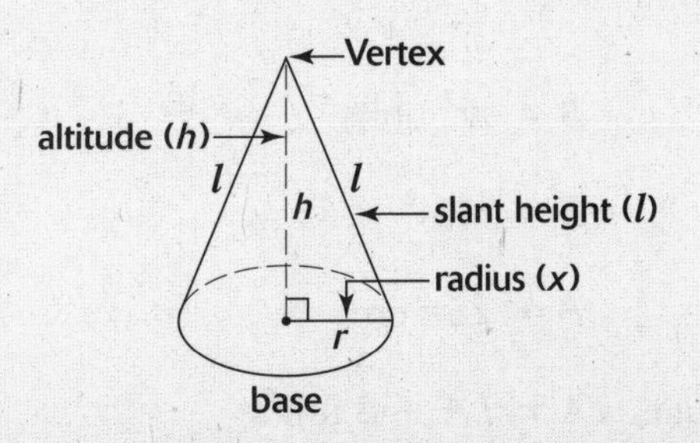

Notice the names of the relevant parts. The theorems for finding a right circular cone's lateral and total areas and its volume should seem quite familiar to you.

Theorem 94: The lateral area, LA, of a right circular cone with a base circumference of C and slant height l is given by the equation, $LA_{cone} = \frac{1}{2}Cl$ units^2, or

$$LA_{cone} = \frac{1}{2}(2)(\pi)(r)(l) \text{ units}^2 \text{ or}$$
$$LA_{cone} = \pi rl$$

Theorem 95: The total area, TA, of a right circular cone with lateral area, LA, and a base area B is given by the equation,

$$TA_{cone} = LA + B \text{ units}^2 \text{ or}$$
$$TA_{cone} = \pi rl + \pi r^2 \text{ units}^2 \text{ or}$$
$$TA_{cone} = \pi r(l + r) \text{ units}^2$$

Theorem 96: The volume of a right circular cone, V, with base area B and altitude h is given by the equation,

$$V_{cone} = \frac{1}{3}(B)(h) \text{ units}^3 \text{ or}$$
$$V_{cone} = \frac{1}{3}(\pi)(r^2)(h) \text{ units}^3$$

Example Problems

These problems show the answers and solutions. All three problems refer to a right circular cone with $r = 5$, $h = 6$, and $l = 9$ centimeters.

1. Find the lateral area of the right circular cone.

***Answer:* 45π cm²**

Write the LA formula: $LA = \pi rl$ units^2

Substitute: $LA = \pi(5)(9)$

Multiply once: $LA = \pi(45)$

. . . and that's it: $LA = 45\pi$ cm^2

2. Find the total area of the right circular cone.

***Answer:* 70π cm²**

First find B: $B = \pi r^2$ units^2

Substitute: $B = \pi(5)^2$

Multiply: $B = 25\pi$ cm^2

Now write the TA formula: $TA = LA + B$ units^2

Substitute: $TA = 45\pi + 25\pi$

Add: $TA = 70\pi$ cm^2

3. Find the volume of the right circular cone.

***Answer:* 50π cm³**

Write the V formula: $V = \frac{1}{3}(B)(h)$ units³

Substitute (from previous): $V = \frac{1}{3}(25\pi)(6)$

Multiply by $\frac{1}{3}$: $V = (25\pi)(2)$

Multiply: $V = 50\pi$ cm³

Was that not cool? Just one more solid is needed to round out the chapter.

Spheres

The **sphere** is the solid equivalent of the circle. It is the set of all points in space equidistant from a fixed point, its center. This distance is the sphere's **radius**. Since a sphere doesn't have any bases or sides, it has no base area or lateral area. It does, however, have surface area—the area of its skin—and, of course, it has volume.

> **Theorem 97:** The surface area of a sphere, S, may be found by the equation $S = 4\pi r^2$ units².
> **Theorem 98:** The volume of a sphere, V, with radius r is given by the equation,

$$V_{\text{sphere}} = \frac{4}{3}(\pi r^3) \text{ units}^3$$

O r

The preceding figure represents a sphere of radius r.

Example Problems

These problems show the answers and solutions.

1. Find the surface area of a sphere of radius 6 inches.

***Answer:* 144π in²**

First write the formula: $S = 4\pi r^2$ units²

Substitute: $S = 4\pi(6)^2$

Square the 6: $S = 4\pi(36)$

Multiply: $S = 144\pi$ in²

2. Find the volume of a sphere of radius 9 mm.

***Answer:* 972π mm³**

First write the formula: $V = \frac{4}{3}(\pi r^3)$ units^3

Substitute: $V = \frac{4}{3}(\pi)(9^3)$ units^3

Cube the 9: $V = \frac{4}{3}(729)\pi$

Simplify by factoring out 3: $V = 4(243)\pi$

Multiply: $V = 972\pi$ mm^3

3. A sphere has a surface area of 324π square centimeters. What is its radius?

***Answer:* 9 cm**

First write the formula: $S = 4\pi r^2$ units^2

Substitute: $324\pi = 4\pi r^2$

Divide both sides by 4π: $81 = r^2$

Find the square root of each side: $r = 9$ cm

Work Problems

Use these problems to give yourself additional practice.

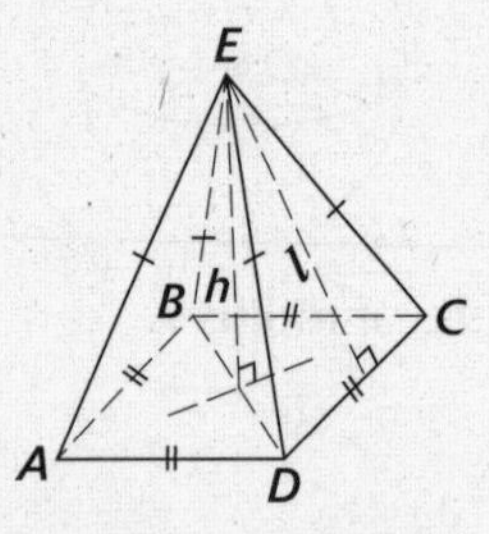

For the preceding figure, $AD = 12$, $DE = 10$, $h = 8$, and $l = 9$ inches.

1. Find the lateral area of the regular square pyramid.

2. Find the volume of the regular square pyramid.

Problems 3–6 refer to a right circular cone with $r = 6$, $h = 6$, and $l = 8$ centimeters.

3. Find the total area of the right circular cone with the given dimensions.

4. Find the volume of the right circular cone with the given dimensions.

5. Find the surface area of a sphere of radius 8 inches.

6. Find the volume of a sphere of radius 12 mm.

Worked Solutions

1. **216 in²**

The perimeter of the base, p, is found by multiplying $4 \times 12 = 48$.

Write the LA formula: $LA = \frac{1}{2}pl$ units^2

Substitute: $LA = \frac{1}{2}(48)(9)$

Multiply by $\frac{1}{2}$: $LA = 24(9)$

Multiply: $LA = 216$ in^2

2. **384 in³**

Use the V formula: $V = \frac{1}{3}Bh$ units^3

Substitute: $V = \frac{1}{3}(12)(12)(8)$

Multiply (12)(12) $V = \frac{1}{3}(144)(8)$

Multiply 144 by $\frac{1}{3}$: $V = (48)(8)$

Multiply: $V = 384$ in^3

3. **84π cm²**

First find B: $B = \pi r^2$ units^2

Substitute: $B = \pi(6)^2$

Multiply: $B = 36\pi$ cm^2

Now write the TA formula: $TA = \pi rl + B$ units^2

Substitute: $TA = (6)(8)\pi + 36\pi$

Multiply: $TA = 48\pi + 36\pi$

Add: $TA = 84\pi$ cm^2

4. **72π cm³**

Write the V formula: $V = \frac{1}{3}(B)(h)$ units^3

Substitute (from previous): $V = \frac{1}{3}(36\pi)(6)$

Multiply by $\frac{1}{3}$: $V = (36\pi)(2)$

Multiply: $V = 72\pi$ cm^3

5. 256π in^2

First write the formula: $S = 4\pi r^2$ units^2

Substitute: $S = 4\pi(8)^2$

Square the 6: $S = 4\pi(64)$

Multiply: $S = 256\pi$ in^2

6. 2304π mm^3

First write the formula: $V = \frac{4}{3}(\pi r^3)$ units^3

Substitute: $V = \frac{4}{3}(\pi)(12)^3$

Cube the 12: $V = \frac{4}{3}(1728)\pi$

Simplify by factoring out 3: $V = 4(576)\pi$

Multiply: $V = 2304\pi$ mm^3

Chapter 10

Coordinate Geometry

Rene Descartes, a French mathematician/philosopher of the seventeenth century discovered that it is possible to assign a set of three coordinates to every point in space, so the system often is referred to as Cartesian [pronounced Kar-**tee**-zeeun] coordinates. Luckily for you, we're not dealing with the three-dimensional system, but only plane (or rectangular) coordinates. The focal point of the **rectangular coordinate system** is a pair of perpendicular axes, labeled x and y.

Locating Points on Coordinate Axes

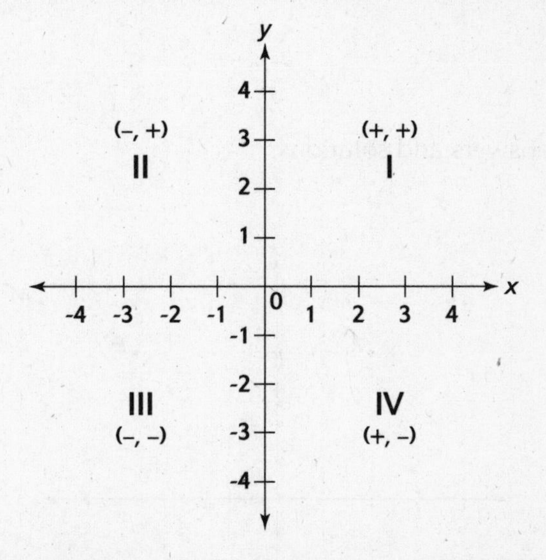

As you can see from the preceding figure, the horizontal axis is labeled x and the vertical one y. Each point in a plane may be assigned a pair of coordinates, in the order, (x, y), also known as an **ordered pair**. The point where the axes cross is called **the origin** and has coordinates (0, 0). Moving horizontally along the x-axis, each step to the right is positive; each step to the left is negative. In the y direction, up is positive and down is negative. Notice that the four **quadrants** (quarters) created by the axes are labeled with Roman numerals and proceed counterclockwise around the origin. The symbols in parentheses indicate the sign attached to all (x, y) coordinates that are in each quadrant. To reinforce that idea, check out the axes in the following figure.

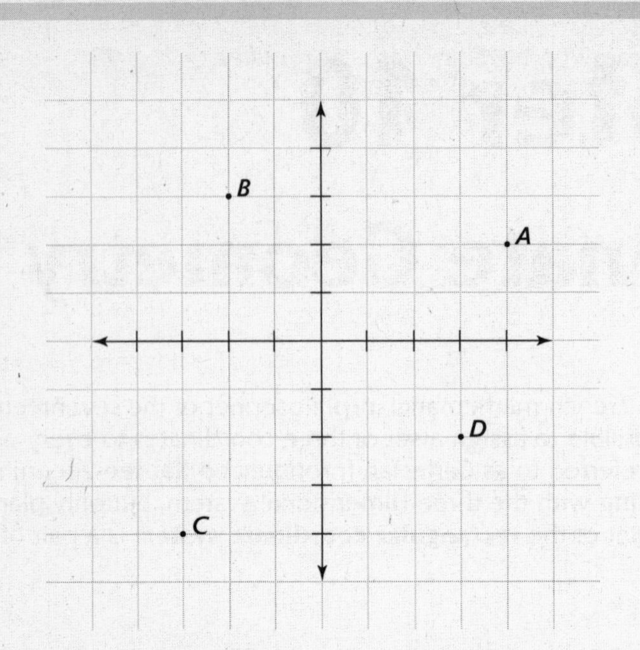

The coordinates of point *A* are (4, 2). Remember, it's horizontal first, then vertical. The coordinates of point *B* are (−2, 3). *C* is at (−3, −4), and *D* is at (3, −2).

Example Problems

These problems show the answers and solutions.

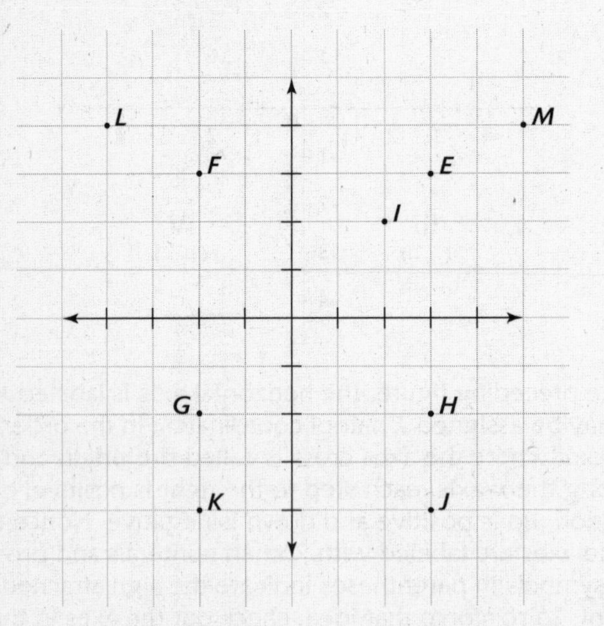

1. What are the coordinates of *F* and *H*?

   ***Answer:* F: (−2, 3); *H*: (3, −2)** To get to *F* from the origin, we went left 2 (that's negative 2) and up 3 (that's positive 3). Since the horizontal coordinate goes first, that's the ordered pair: (−2, 3). The positive sign is understood, and so by convention is not written. To get to *H* from the origin, we went right 3 (that's positive 3), and down 2 (that's negative 2). Since the horizontal coordinate goes first, that's the ordered pair: (3, −2).

2. a) What is the name of the point with coordinates (−2, −4)?

b) What is the name of the point with coordinates (−4, 4)?

Answer:* a) *K*, b) *L Coordinates (−2, −4) tell us to start at the origin, go left two and then down four. That puts us at point *K*.

Coordinates (−4, 4) tell us to start at the origin, go left four and then up four. That puts us at point *L*.

3. Find the distance from *E* to *H* (*EH*).

***Answer:* 5** By Postulate 7 (the *Ruler Postulate*), each point on a line or a segment can be paired with a single real number, which is known as that point's **coordinate**. The distance between two points is the absolute value of the difference of their coordinates.

$EH = 3 - (-2)$

$EH = 5$

The Distance Formula

In the figure that follows, *A* is at (3, 3), *B* is at (3, −3), and *C* is at (−5, −3). To find *AB* and *BC* is easy. It's just a matter of subtraction:

$AB = 3 - (-3)$ $BC = 3 - (-5)$

$AB = 6$ $BC = 8$

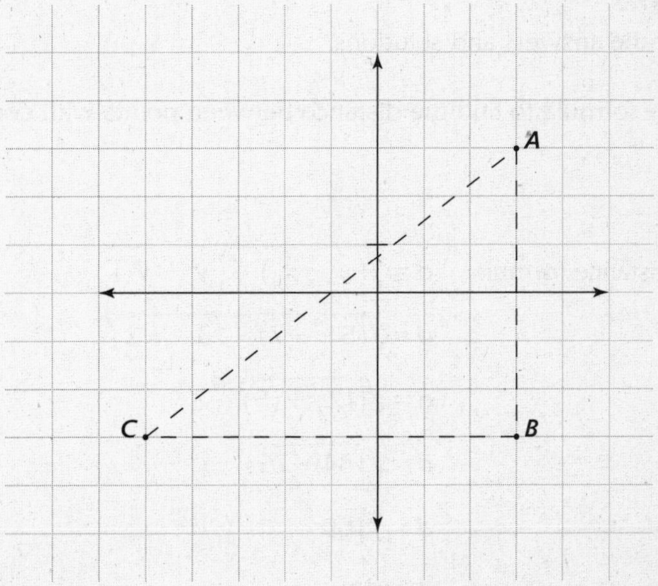

A
C
B

To find *AC*, simple subtraction won't do. We have to notice that $\triangle ABC$ is a right triangle with $\overline{AC}$ as the hypotenuse. Using the side lengths we've already computed, we can find *AC* by means of the Pythagorean Theorem:

$AC^2 = AB^2 + BC^2$

$AC = \sqrt{AB^2 + BC^2}$
$AC = \sqrt{6^2 + 8^2}$
$AC = \sqrt{36 + 64}$
$AC = \sqrt{100}$
$AC = 10$

All the information presented in this section up to this point has been to show how the distance formula is derived. If the coordinates of point C are (x_1, y_1), and the coordinates of point A are (x_2, y_2), then $BC = |x_2, - x_1|$, and $AB = |y_2, - y_1|$. Applying this logic to the second step of the solution to finding AC, then we find:

$$AC = \sqrt{(x_2 - x_1)^2 + (y_2 - y_1)^2}$$

This is nicely summarized in Theorem 94:

> **Theorem 94** (*The Distance Formula*): If the coordinates of two points are (x_1, y_1) and (x_2, y_2), then the distance, d, between the two points is expressed by the formula:
>
> $$d = \sqrt{(x_2 - x_1)^2 + (y_2 - y_1)^2}$$

Notice that it's just another form of the Pythagorean Theorem. It doesn't matter which point you select as point 2, as long as you select the same point 2 for both the x and y values.

Example Problems

These problems show the answers and solutions.

1. Use the distance formula to find the distance between points with coordinates (5, 3) and (−7, −2).

   ***Answer:* 13**

   Start with the distance formula: $d = \sqrt{(x_2 - x_1)^2 + (y_2 - y_1)^2}$

   Substitute: $d = \sqrt{(5 - -7)^2 + (3 - -2)^2}$

   Subtract: $d = \sqrt{(12)^2 + (5)^2}$

   Square: $d = \sqrt{144 + 25}$

   Add: $d = \sqrt{169}$

   Take the square root: $d = 13$

2. $\triangle ABC$ has vertices A (4, 4), B (14, −4), and C (−6, −4). Show that the triangle is isosceles.

   ***Answer:* $AB = AC$**

Since the y-coordinates of B and C are both -4, we can find the length of BC by subtracting its x-coordinates:

$$BC = 14 - -6 = 20$$

The lengths of AB and AC are found by using the distance formula. Below, the instructions are in the center column, and apply to both AB on the left, and AC on the right.

$$AB = \sqrt{(x_2 - x_1)^2 + (y_2 - y_1)^2}$$

First, the formula $AC = \sqrt{(x_2 - x_1)^2 + (y_2 - y_1)^2}$

$AB = \sqrt{(14-4)^2 + (-4-4)^2}$	Substitute	$AC = \sqrt{(4 - -6)^2 + (4 - -4)^2}$
$AB = \sqrt{(10)^2 + (-8)^2}$	Subtract	$AC = \sqrt{(10)^2 + (8)^2}$
$AB = \sqrt{100 + 64}$	Square	$AC = \sqrt{100 + 64}$
$AB = \sqrt{164}$	Add	$AC = \sqrt{164}$
$AB = \sqrt{4 \times (41)}$	Simplify	$AC = \sqrt{4 \times (41)}$
$AB = 2\sqrt{41}$	Simplify	$AC = 2\sqrt{41}$

$AB = AC$, so the triangle is isosceles.

The Midpoint Formula

Given the coordinates of the endpoints of a line segment, the midpoint may be thought of as being the average of those endpoints. That should help you to remember the midpoint formula, since you should be used to adding two numbers and dividing by two to find their average. As you might have expected, there is a theorem to cover this subject:

Theorem 95 (*The Midpoint Formula*): If the coordinates of two endpoints of a segment are given by (x_1, y_1) and (x_2, y_2), respectively, then the midpoint of that segment, M, may be found by the formula:

$$M = \left(\frac{x_1 + x_2}{2}, \frac{y_1 + y_2}{2}\right)$$

Example Problems

These problems show the answers and solutions.

1. Segment PQ has endpoints $P(-5, 3)$ and $Q(7, -3)$. Find the coordinates of its midpoint, R.

   Answer: **(1, 0)**

First, write the formula: $M=\left(\frac{x_1+x_2}{2},\frac{y_1+y_2}{2}\right)$

Substitute: $M=\left(\frac{-5+7}{2},\frac{3+-3}{2}\right)$

Add in numerators: $M=\left(\frac{2}{2},\frac{0}{2}\right)$

Divide: $M = (1, 0)$

2. Use the distance formula to prove that you actually found the midpoint in Problem 1.

***Answer:* It is proven.** (Below, the instructions are in the center column, and apply to both *MP* on the left, and *MQ* on the right.)

$MP=\sqrt{(x_2-x_1)^2+(y_2-y_1)^2}$	First, the formula.	$MQ=\sqrt{(x_2-x_1)^2+(y_2-y_1)^2}$
$MP=\sqrt{(-5-1)^2+(3-0)^2}$	Substitute.	$MQ=\sqrt{(7-1)^2+(-3-0)^2}$
$MP=\sqrt{(-6)^2+(3)^2}$	Subtract.	$MQ=\sqrt{(6)^2+(-3)^2}$

We can really see that they're the same here, but . . .

$MP=\sqrt{36+9}$	Square	$MQ=\sqrt{36+9}$
$MP=\sqrt{45}$	Add	$MQ=\sqrt{45}$
$MP=\sqrt{9\times5}$	Simplify	$MQ=\sqrt{9\times5}$
$MP=3\sqrt{5}$	Simplify	$MQ=3\sqrt{5}$

MP = *MQ*, so *M* is the midpoint of *PQ*.

3. Segment *AB* has midpoint *D* (−1, 1), and endpoint *A* (−6, 7). Find the coordinates of *B*.

***Answer:* (4, −5)**

Start with the formula: $M=\left(\frac{x_1+x_2}{2},\frac{y_1+y_2}{2}\right)$

Substitute: $(-1,1)=\left(\frac{-6+x_2}{2},\frac{7+y_2}{2}\right)$

(Below, the instructions are in the center column, and apply to both *x*-coordinates on the left, and *y*-coordinates on the right.)

$\frac{-6+x_2}{2}=-1$	Separate *x*'s and *y*'s:	$\frac{7+y_2}{2}=1$
$-6+x_2=-2$	Multiply by 2:	$7+y_2=2$
$x_2=-2+6$	Collect terms:	$y_2=2-7$
$x_2=4$	Combine constants.	$y_2=-5$

B's coordinates are (4, −5).

Work Problems

Use these problems to give yourself additional practice.

1. Rectangle *ABCD* has vertices *A* (−10, 4), *B* (14, 4), *C* (14, −6), and *D* (−10, −6). Find the length of diagonal segment *AC*.
2. The diameter of a circle has endpoints at (−2, 8) and (10, −12). What are the coordinates of the circle's center?
3. What is special about a line segment with endpoints (25, −13) and (25, 22)?
4. What is special about a line segment with endpoints (−18, −13) and (25, −13)?
5. Find the length of a line segment with endpoints *C* (24, −6), and *D* (−12, 14).

Worked Solutions

1. **26**

Start with the distance formula: $d=\sqrt{(x_2-x_1)^2+(y_2-y_1)^2}$

Substitute: $d=\sqrt{(14--10)^2+(-6-4)^2}$

Subtract: $d=\sqrt{(24)^2+(-10)^2}$

Square: $d=\sqrt{576+100}$

Add: $d=\sqrt{676}$

Take the square root: $d = 26$

2. **(4, −2)**

First, write the formula: $M=\left(\frac{x_1+x_2}{2},\frac{y_1+y_2}{2}\right)$

Substitute: $M=\left(\frac{-2+10}{2},\frac{8+-12}{2}\right)$

Add figures in the numerators: $M=\left(\frac{8}{2},\frac{-4}{2}\right)$

Divide: $M = (4, -2)$

3. **It is vertical.** Any two or more points with the same *x*-coordinate lie on a vertical line.

4. **It is horizontal.** Any two or more points with the same *y*-coordinate lie on a horizontal line.

5. $4\sqrt{106}$

Use the distance formula:	$d=\sqrt{(x_2-x_1)^2+(y_2-y_1)^2}$
Substitute:	$d=\sqrt{(14--6)^2+(-12-24)^2}$
Subtract:	$d=\sqrt{(20)^2+(-36)^2}$
Square:	$d=\sqrt{400+1296}$
Add:	$d=\sqrt{1696}$
Simplify the square root:	$d=\sqrt{16\times 106}$
Remove the 4 from the bracket:	$d=4\sqrt{106}$

Slope of a Line

This topic is not the slippery slope it's often made out to be. The **slope of a line** is the measure of its steepness. It often is referred to as rise over run, and that's a good way to think of it. In fact, it's the difference of rise over the difference in run, or $\frac{dy}{dx}$.

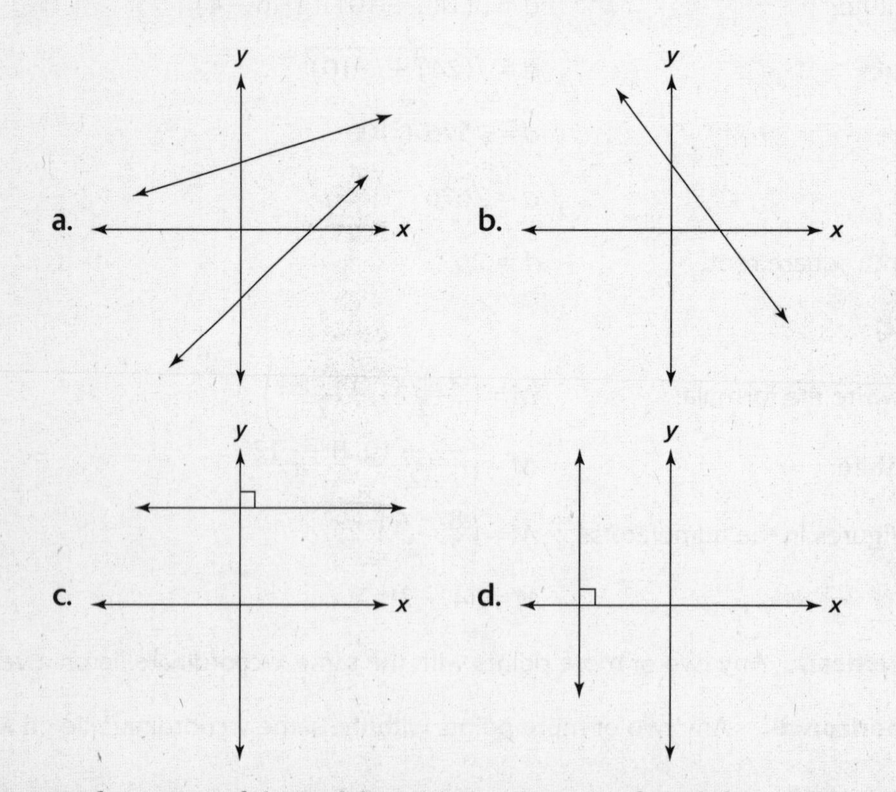

Basically, there are four types of slope: A line rising as it moves from left to right, such as in *a*, has a positive slope. The line in *b* is moving downward from left to right, and so has a negative slope. The slope of any horizontal line, as in *c*, or the *x*-axis, is 0. The line in *d*, as well as the *y*-axis, has a slope that is considered undefined. The letter *m* usually is used to stand for slope, so given line *AB* with coordinates $A\ (x_1,\ y_1)$, and $B\ (x_2, y_2)$, then

$$m = \frac{y_2 - y_1}{x_2 - x_1}, \text{ as long as } x_2 \neq x_1$$

Note that *A* and *B* cannot be on a vertical line, so x_1 can't be equal to x_2. If $x_1 = x_2$, then the line is vertical, and so its slope is undefined.

Example Problems

These problems show the answers and solutions. The problems that follow all refer to this diagram.

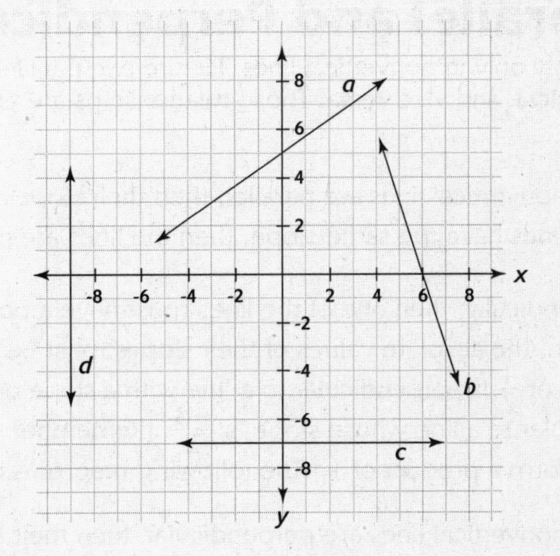

1. Find the slope of line *a*.

Answer: $\frac{2}{3}$ (−3, 3) and (0, 5) are on line *a*. Let's use them to substitute into the slope formula:

First the formula: $m = \frac{(y_2 - y_1)}{(x_2 - x_1)}$

Substitute: $m = \frac{(5 - 3)}{(0 - -3)}$

Subtract: $m = \frac{2}{3}$

The slope is $\frac{2}{3}$

2. Find the slope of line *b*.

Answer: **−3** (5, 3) and (6, 0) are on line *b*. Let's use them to substitute into the slope formula:

First the formula: $m = \frac{(y_2 - y_1)}{(x_2 - x_1)}$

Substitute: $m = \frac{(0 - 3)}{(6 - 5)}$

Subtract: $m = \frac{(-3)}{1}$

The slope is −3

3. Find the slope of line c.

   ***Answer:* 0** The slope of any horizontal line is zero. Think about it: If slope is rise over run, no matter how far it runs, its rise is 0.

4. Find the slope of line d.

   ***Answer:* Undefined** The slope of any vertical line is undefined.

Slopes of Parallel and Perpendicular Lines

All rules in this section apply only to nonvertical lines. Picture two parallel lines. If lines are parallel, their slopes must be identical, and vice versa. Those relationships are expressed in the following two theorems:

> **Theorem 96:** If two nonvertical lines are parallel, then their slopes are the same.
>
> **Theorem 97:** If two lines have the same slope, then the lines are parallel.

When two lines are perpendicular, then one of the lines must have a positive slope and the other a negative one. In addition, the absolute values of their slopes must be reciprocals. In other words, a line with a slope of $\frac{1}{2}$ is perpendicular to a line with a slope of -2. Similarly, a line with a slope of $\frac{2}{3}$ is perpendicular to a line whose slope is $\frac{-3}{2}$. (Remember, reciprocals are numbers that multiply together to form a product of 1.) The following theorems cover this.

> **Theorem 98:** If two nonvertical lines are perpendicular, then their slopes are negative reciprocals of one another (i.e., the product of the slopes is -1).
>
> **Theorem 99:** If the slopes of two lines multiply together to form a product of -1, then the lines are perpendicular and nonvertical.

Since horizontal and vertical lines are always perpendicular, we could add a theorem to say that if one line has a slope of zero and the other has a slope that is undefined, then the lines are perpendicular, but why waste the space?

Example Problems

These problems show the answers and solutions.

1. A line has a slope of $\frac{3}{4}$. For a line to be parallel to it, that line must have a slope of _____?

   Answer: $\frac{3}{4}$ By Theorem 96, parallel lines have the same slope.

2. A line has a slope of $\frac{3}{4}$. For a line to be perpendicular to it, that line must have a slope of _____?

   Answer: $\frac{-4}{3}$ By Theorem 98, if two nonvertical lines are perpendicular, then the product of their slopes is -1.

   $$\frac{3}{4}\times\frac{-4}{3}=-1$$

3. Two lines are perpendicular. One has a slope of 0. What is the slope of the other?

***Answer:* Undefined** A line with a slope of 0 is horizontal, therefore its perpendicular must be vertical. A vertical line's slope is undefined.

Work Problems

Use these problems to give yourself additional practice. All problems refer to the following graph.

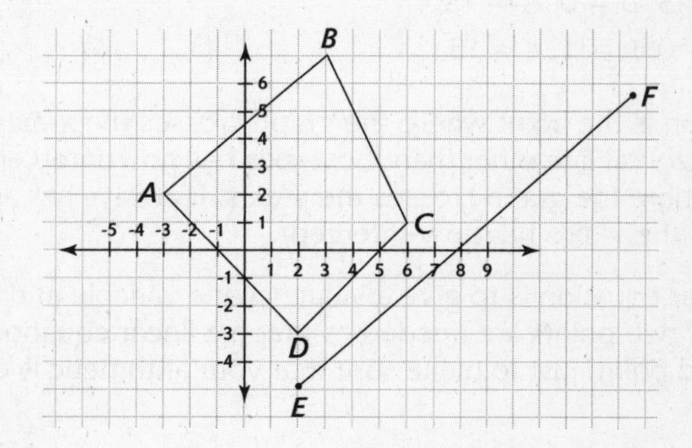

1. Find the slope of $\overline{CD}$.
2. Are *AB* and *CD* parallel?
3. Tell why *AD* and *CD* are or are not perpendicular.
4. Find the slope of *BC*.
5. Does *EF* have a relationship to any other line segment on the graph? If so, how? Prove it.

Worked Solutions

1. **1** *CD* touches the corner of each square as it rises. One up, one over, and so on. That's $\frac{dy}{dx} = \frac{1}{1} = 1$.
2. **No** *AB*'s slope is $\frac{5}{6}$; *CD*'s slope is 1. Parallel lines have the same slope.
3. **Slopes are − reciprocals: *m* for *AD* = −1; *m* for *CD* = 1** $AD \perp CD$.
4. **−2** For every one to the right, *BC* goes down 2. $m = \frac{dy}{dx} = \frac{-2}{1} = -2$
5. ***EF* || *AB*** Both have slopes of $\frac{5}{6}$, and so are parallel.

Equations of Lines

The coordinate plane can be used to graph any equation involving one or two variables. Generally speaking, the following principles govern that graph. If a point lies on the graph of an equation, then the coordinates of that point make the equation a true mathematical statement, and vice versa. A **linear equation** is any equation whose graph is a straight line.

Standard Form of a Linear Equation

Any linear equation can be written in **standard form**, $ax + by = c$, where a, b, and c are real numbers. In addition, a and b may not *both* have a value of 0.

Here are some examples of linear equations in standard form, and their a, b, and c values.

$x + y = 0$	$a = 1, b = 1, c = 0$
$4x - 5y = 11$	$a = 4, b = -5, c = 11$
$3x = -9$	$a = 3, b = 0, c = -9$
$y = 13$	$a = 0, b = 1, c = 13$

The **x-intercept** of a graph is the point where the graph crosses the x-axis. It always has a y-coordinate of 0. A horizontal line other than the x-axis has no x-intercept. The **y-intercept** of a graph is the point where the graph crosses the y-axis. It always has an x-coordinate of 0. A vertical line other than the y-axis has no y-intercept.

One way to graph a linear equation is to give a value to one variable and solve the equation for the other. A minimum of two points are needed to graph a linear equation, but it is usually a good idea to have a third point, just to make sure that your arithmetic is correct.

Example Problems

These problems show the answers and solutions.

1. Substitute values for x to get y values of at least 3 points. Then graph the equation $2x + 3y = 6$.

Answer:

x	y
-3	4
0	2
3	0

First, rewrite the equation to be solved for y: $3y = 6 - 2x$.

Then, substitute values for x: If $x = -3$, then $y = 4$, per the table.

Answer:

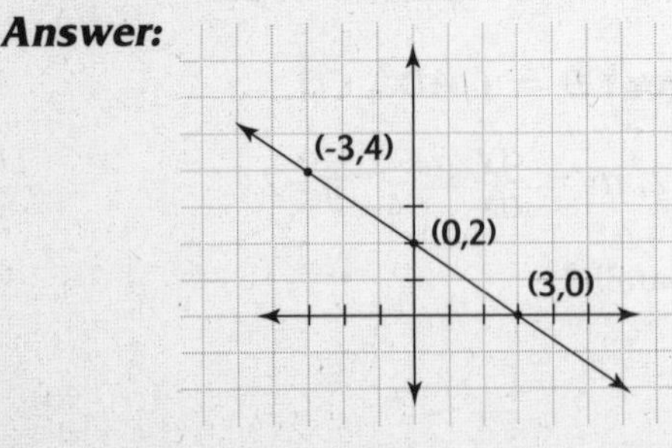

It is good practice to substitute both a positive and a negative value, as well as 0. Why did you choose −3 and 3 rather than 1 or 2? Because they would have solved as fractional values for y. It is difficult to plot points with fractional coordinates. Finally, plot the sets of coordinates on the axes and draw the line that passes through them. If the three points had not fallen on a straight line, you would have to go back and check your computation.

2. Find the x- and y-intercepts for the equation, $2x + 3y = 18$.

Answer: **(9, 0), (0, 6)**

To find the y-intercept: Let $x = 0$

Then $2(0) + 3y = 18$

$3y = 18$

$y = 6$

To find the x-intercept: Let $y = 0$

Then $2x + 3(0) = 18$

$2x = 18$

$x = 9$

3. Use the x- and y-intercepts found in problem 2 to draw the graph of $2x + 3y = 18$.

Answer:

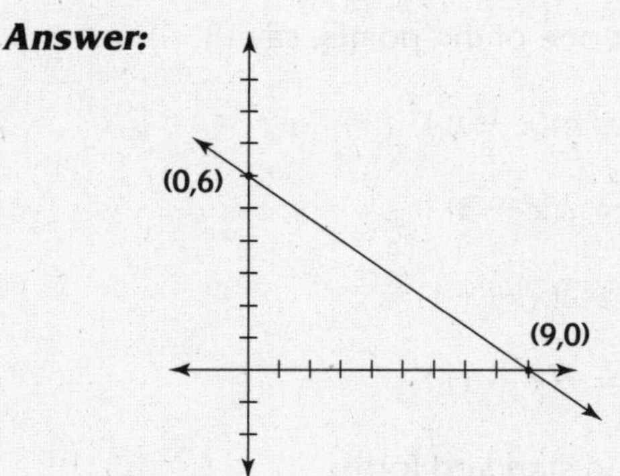

Just plot the points (0, 6) and (9, 0) and draw the line on which they lie.

Point-Slope Form of a Linear Equation

Suppose that A is a point on a grid with coordinates (x_1, y_1), and B, on the same grid, has coordinates (x, y). Then the slope, m, of the line through A and B is represented by this equation:

$$m = \frac{y - y_1}{x - x_1}$$

By applying the cross products property, we'll get $y - y_1 = m(x - x_1)$. This is called the point-slope form of a line. Of course, it does not apply to a vertical line.

> **Theorem 100:** The point-slope form of a nonvertical line passing through (x, y) and having slope m is $y - y_1 = m(x - x_1)$.

This form is useful for finding the equation of a line when what you have is two points on that line.

Slope-*y*-intercept Form of a Linear Equation

Should you happen to know the slope of a line and its *y*-intercept, you can write the equation of the line using the slope-*y*-intercept form, as described in the last theorem in this book:

Theorem 101: The slope-*y*-intercept form of a nonvertical line with slope m and *y*-intercept vertical coordinate b is $y = mx + b$.

Example Problems

These problems show the answers and solutions.

1. Find the equation of a line containing the points (3, 4) and (−1, −4) and write it in the point-slope form.

***Answer:* $y - 4 = 2(x - 3)$ or $y + 4 = 2(x + 1)$**

First, find the slope: $m = \frac{y - y_1}{x - x_1}$

Substitute: $m = \frac{4 - -4}{3 - -1}$ $m = \frac{8}{4}$ $m = 2$

Now move to the point-slope formula and pick one of the points, say (3, 4).

Start with the formula: $y - y_1 = m(x - x_1)$

Substitute: $y - 4 = 2(x - 3)$

For the other point (−1, −4), we get $y - -4 = 2(x - -1)$

Subtracting, we get: $y + 4 = 2(x + 1)$

2. Take the equation from problem 1 and write it in standard form.

***Answer:* $2x - y = 2$**

Standard form, you may recall, is $ax + by = c$

Take the point-slope form and collect terms: $y + 4 = 2(x + 1)$

Clear parentheses: $y + 4 = 2x + 2$

Get the x and y terms together: $4 = 2x - y + 2$

. . . and collect constants: $2 = 2x - y$

Lastly, turn it around: $2x - y = 2$

3. Finally, rewrite the same equation in slope-y-intercept form.

Answer: $\mathbf{y = 2x - 2}$

The form is: $y = mx + b$

Start with standard form: $2x - y = 2$

Get the y alone on the left: $-y = -2x + 2$

Divide through by -1: $y = 2x - 2$

Work Problems

Use these problems to give yourself additional practice.

1. Find the slope and y-intercept of the line with equation $6x - 8y = 40$.
2. The equations of three lines are as follows: l_1 $6x + 15y = 30$, l_2 $2x + 5y = 15$, and l_3 $5x - 2y = 4$. Which, if any, are parallel or perpendicular?
3. Find the point-slope form for the equation of the line passing through (4, 4) and (9, −2).
4. Find an equation of the line passing through (3, 3) and perpendicular to the equation in problem 1.
5. Express the solution from problem 4 in standard form for a linear equation.

Worked Solutions

1. $\mathbf{\frac{3}{4}}$**; (0, −5)**

Start with: $6x - 8y = 40$

First, put into slope-y-intercept format: $-8y = -6x + 40$

Divide both sides by -8: $y = \frac{6}{8}x + -5$

Simplify: $y = \frac{3}{4}x - 5$

Slope is m, and y-intercept is b in $y = mx + b$

That means the slope is $\frac{3}{4}$; the y-intercept is (0, −5)

2. $\mathbf{l_1 \parallel l_2; l_3 \perp l_1}$ **and** $\mathbf{l_2}$ All three equations must be put into $y = mx + b$ form.

l_1 $6x + 15y = 30$,

$15y = -6x + 30$

Divide by 15: $y = \frac{-6}{15}x + 2$

Simplify: $y = \frac{-2}{5}x + 2$

Slope $l_1 = \frac{-2}{5}$

Next: l_2 $2x + 5y = 15$

$5y = -2x + 15$

Divide by 5: $y = \frac{-2}{5}x + 3$

Slope $l_2 = \frac{-2}{5}$

Finally, l_3 $5x - 2y = 4$

$-2y = -5x + 4$

Divide by −2: $y = \frac{-5}{-2}x + \frac{2}{-2}$

Simplify: $y = \frac{5}{2}x - 1$

Slope $l_3 = \frac{5}{2}$

That makes $l_1 \,||\, l_2$ and l_3 perpendicular to both other lines.

3. $\mathbf{y - 4 = -\frac{6}{5}(x - 4)}$**, or** $\mathbf{y + 2 = \frac{-6}{5}(x - 9)}$

Point-slope form is $y - y_1 = m(x - x_1)$.

First, find m, the slope: $m = \frac{y - y_1}{x - x_1}$

Substitute: $m = \frac{-22 - 4}{9 - 4}$

Subtract: $m = \frac{-6}{5} = -\frac{6}{5}$

Now substitute into the form from either point:

$y - y_1 = m(x - x_1)$ $\quad$ $y - y_1 = m(x - x_1)$

$y - 4 = -\frac{6}{5}(x - 4)$ $\quad$ $y - -2 = -\frac{6}{5}(x - 9)$

$y + 2 = -\frac{6}{5}(x - 9)$

4. $\mathbf{y - 3 = -\frac{4}{3}(x - 3)}$ The slope of the equation in 1 was $\frac{3}{4}$, so you need it's negative reciprocal, for the slope of a perpendicular: $m = -\frac{4}{3}$

Use the point-slope formula, since you have a point and a slope: $y - y_1 = m(x - x_1)$

Substitute: $y - 3 = -\frac{4}{3}(x - 3)$

That's that!

5. $\mathbf{4x + 3y = 21}$ Standard form is $ax + bx = c$.

$y - 3 = -\frac{4}{3}(x - 3)$

To get rid of the fraction, ×3 everything: $3y - 9 = -4(x - 3)$

Multiply by −4 to clear the parentheses: $3y - 9 = -4x - 12$

Collect terms; variables left, constants right: $3y + 4x = 12 + 9$

Arrange variables and combine constants: $4x + 3y = 21$

Customized Full-Length Exam

1. How many letters are required to name a plane?

 Answer: 1

If you answered correctly, go to problem 3.
If you answered incorrectly, go to problem 2.

2. What is the maximum number of lines in a plane that can contain two of the points A, B, and C?

 Answer: 3

If you answered correctly, go to problem 3.
If you answered incorrectly, review "Lines" and "Planes" on pages 37 and 38.

3. If two lines intersect, how many points of intersection are possible?

 Answer: 1

If you answered correctly, go to problem 5.
If you answered incorrectly, go to problem 4.

4. If a point lies outside a line, how many planes can contain both?

 Answer: 1

If you answered correctly, go to problem 5.
If you answered incorrectly, review "Postulates and Theorems" on page 38.

5. What geometric figure is represented by $\overline{ST}$?

 Answer: a line segment

If you answered correctly, go to problem 7.
If you answered incorrectly, go to problem 6.

6. Given figure $\overline{JK}$, what is the meaning of JK?

 Answer: the length of the segment

If you answered correctly, go to problem 7.
If you answered incorrectly, review "Line Segments" on page 41.

7. If N lies between M and P on a line, then what must be true?

 Answer: $MN + NP = MP$

If you answered correctly, go to problem 9.
If you answered incorrectly, go to problem 8.

8. What is the name given to the point equidistant from both ends on a line segment?

Answer: midpoint

If you answered correctly, go to problem 9.
If you answered incorrectly, review "Segment Addition and Midpoint" on page 42.

9. Where is $\angle KPL$ formed?

Answer: At the endpoints of $\overrightarrow{PK}$ and $\overrightarrow{PL}$

If you answered correctly, go to problem 11.
If you answered incorrectly, go to problem 10.

10. What is the maximum number of whole degrees a unique angle may contain?

Answer: 359

If you answered correctly, go to problem 11.
If you answered incorrectly, review "Forming and Naming Angles" on page 45.

11. If $m\angle SUV = 48°$ and $m\angle VUT = 40°$, and the angles share side $\overrightarrow{VU}$, what is the $m\angle SUT$?

Answer: 88°

If you answered correctly, go to problem 13.
If you answered incorrectly, go to problem 12.

12. If $m\angle ABD = 108°$ and $m\angle ABC = 30°$, and the angles share side $\overrightarrow{BC}$, what is the $m\angle CBD$?

Answer: 78°

If you answered correctly, go to problem 13.
If you answered incorrectly, review "The Protractor Postulate and Addition of Angles" on page 46.

13. *FG* bisects right $\angle EFH$. What is the degree measure of $\angle EFG$?

Answer: 45°

If you answered correctly, go to problem 15.
If you answered incorrectly, go to problem 14.

14. How many angle bisectors may a 220° angle have?

Answer: 1

If you answered correctly, go to problem 15.
If you answered incorrectly, review "Angle Bisector" on page 47.

15. What is the degree measure of an obtuse angle?

Answer: greater than 90° and less than 180°

If you answered correctly, go to problem 17.
If you answered incorrectly, go to problem 16.

16. What is the degree measure of a reflex angle?

Answer: greater than 180° and less than 360°

If you answered correctly, go to problem 17.
If you answered incorrectly, review "Angles and Angle Pairs" on page 45.

17. What is the name given to two angles that share a vertex and are separated by a common side?

Answer: adjacent angles

If you answered correctly, go to problem 19.
If you answered incorrectly, go to problem 18.

18. What is the name given to two angles whose degree measure totals 90°?

Answer: complementary angles

If you answered correctly, go to problem 19.
If you answered incorrectly, review "Special Angle Pairs" on page 50.

19. What is the name given to two or more lines that intersect to form 90° angles?

Answer: perpendicular lines

If you answered correctly, go to problem 21.
If you answered incorrectly, go to problem 20.

20. What is the name given to two or more lines in the same plane that never intersect?

Answer: parallel lines

If you answered correctly, go to problem 21.
If you answered incorrectly, review "Special Lines and Segments" on page 53.

21. What is the name given to two angles formed by parallel lines and their transversal that are in the same relative positions on each of the parallel lines?

Answer: corresponding angles

If you answered correctly, go to problem 23.
If you answered incorrectly, go to problem 22.

22. What do we call angles between the parallel lines and on opposite sides of the transversal?

Answer: alternate interior angles

If you answered correctly, go to problem 23.
If you answered incorrectly, review "Angles Created by Lines and a Transversal" on page 55.

23. If parallel lines are cut by a transversal, which three sets of angles are equal?

Answer: alternate interior angles, alternate exterior angles, corresponding angles

If you answered correctly, go to problem 25.
If you answered incorrectly, go to problem 24.

24. If parallel lines are cut by a transversal, what is true of their consecutive interior angles and their consecutive exterior angles?

Answer: They are supplementary.

If you answered correctly, go to problem 25.
If you answered incorrectly, review "Angles Created by Parallel Lines and a Transversal" on page 56.

25. What kind of triangle has two angles of equal measure?

Answer: isosceles

If you answered correctly, go to problem 27.
If you answered incorrectly, go to problem 26.

26. What kind of triangle has one angle greater than 90°?

Answer: an obtuse triangle

If you answered correctly, go to problem 27.
If you answered incorrectly, review "Classifying Triangles by Angles" on page 65.

27. What is the name given to the equal sides of an isosceles triangle?

Answer: legs

If you answered correctly, go to problem 29.
If you answered incorrectly, go to problem 28.

28. What is the name given to the sides of a right triangle that form the right angle?

Answer: legs

If you answered correctly, go to problem 29.
If you answered incorrectly, review "Specially Named Sides and Angles" on page 68.

29. How many altitudes of an obtuse triangle fall inside the triangle?

Answer: 1

If you answered correctly, go to problem 31.
If you answered incorrectly, go to problem 30.

30. How many altitudes of a right triangle is (are) neither inside nor outside the triangle?

Answer: 2

If you answered correctly, go to problem 31.
If you answered incorrectly, review "Base and Altitude" on page 70.

31. In a triangle, what is the name given to line segments drawn from a vertex to the midpoint of the opposite side?

Answer: median

If you answered correctly, go to problem 33.
If you answered incorrectly, go to problem 32.

32. In a triangle, how many medians may be drawn?

Answer: 3

If you answered correctly, go to problem 33.
If you answered incorrectly, review "Median" on page 71.

33. Which of these is *not* a reason for proving triangles congruent: SSS, SSA, or AAS?

Answer: SSA

If you answered correctly, go to problem 35.
If you answered incorrectly, go to problem 34.

34. Which of these is *not* a reason for proving triangles congruent: ASA, SAA, or HL?

Answer: none

If you answered correctly, go to problem 35.
If you answered incorrectly, review "Proofs of Congruence" on page 73.

35. $\triangle ABC \cong \triangle DEF$ in that order. Why does $m\angle B = m\angle E$?

Answer: CPCTC

If you answered correctly, go to problem 37.
If you answered incorrectly, go to problem 36.

36. $\triangle ABC \cong \triangle DEF$ in that order. What is congruent to $\overline{AC}$?

Answer: $\overline{DF}$

If you answered correctly, go to problem 37.
If you answered incorrectly, review "Corresponding Parts (CPCTC)" on page 79.

37. In $\triangle PQR$, $PQ > PR$. Which is greater: $m\angle Q$ or $m\angle R$?

Answer: $m\angle R$

If you answered correctly, go to problem 39.
If you answered incorrectly, go to problem 38.

38. In $\triangle EFG$, $EF = 6$ cm, $FG = 7$ cm. What is the maximum possible length of $\overline{EG}$?

Answer: <13 cm

If you answered correctly, go to problem 39.
If you answered incorrectly, review "Triangle Inequality Theorems" on page 82.

39. What is the sum of the interior angles of a regular octagon?

Answer: 1080°

If you answered correctly, go to problem 41.
If you answered incorrectly, go to problem 40.

40. What is the sum of the exterior angles of a regular octagon?

Answer: 360

If you answered correctly, go to problem 41.
If you answered incorrectly, review "Angle Sums" on page 89.

41. What is a parallelogram with one right angle?

Answer: a rectangle

If you answered correctly, go to problem 43.
If you answered incorrectly, go to problem 42.

42. What is an equilateral parallelogram?

Answer: a rhombus

If you answered correctly, go to problem 43.
If you answered incorrectly, review "Quadrilaterals" on page 92.

43. What do we call a quadrilateral whose diagonals bisect each other?

Answer: a parallelogram

If you answered correctly, go to problem 45.
If you answered incorrectly, go to problem 44.

44. What do we call a quadrilateral with one pair of opposite sides both equal and parallel?

Answer: a parallelogram

If you answered correctly, go to problem 45.
If you answered incorrectly, review "Proofs of Parallelograms" on page 95.

45. What do we call a quadrilateral whose diagonals bisect the opposite angles?

Answer: a rhombus

If you answered correctly, go to problem 47.
If you answered incorrectly, go to problem 46.

46. What do we call a quadrilateral whose diagonals are perpendicular to one another?

Answer: a rhombus

If you answered correctly, go to problem 47.
If you answered incorrectly, review "Rhombus" on page 99.

47. A trapezoid's bases are 15 and 23 inches long, respectively. How long is the trapezoid's median?

Answer: 19 inches

If you answered correctly, go to problem 49.
If you answered incorrectly, go to problem 48.

48. A trapezoid's bases are 22 and 34 cm long, respectively. How long is the trapezoid's median?

Answer: 28 cm

If you answered correctly, go to problem 49.
If you answered incorrectly, review "Special Trapezoids" on page 101.

49. In $\triangle ABC$, $AB = 12$, $BC = 14$, and $AC = 16$. How long is the line joining the midpoints of AB and BC?

Answer: 8

If you answered correctly, go to problem 51.
If you answered incorrectly, go to problem 50.

50. In $\triangle FGH$, $FG = 13$, $GH = 17$, and $HF = 19$. How long is the line joining the midpoints of GH and FH?

Answer: 6.5

If you answered correctly, go to problem 51.
If you answered incorrectly, review "The Midpoint Theorem" on page 102.

51. Find the perimeter of a rectangle that is 6 cm wide and 9 cm long.

Answer: 30 cm

If you answered correctly, go to problem 53.
If you answered incorrectly, go to problem 52.

52. Find the perimeter of a rectangle that is 20 in. wide and 30 in. long.

Answer: 100 in.

If you answered correctly, go to problem 53.
If you answered incorrectly, review "Finding the Perimeter" in "Squares and Rectangles" on page 105.

53. A square has a side 14 inches long. What is its perimeter?

Answer: 56 in.

If you answered correctly, go to problem 55.
If you answered incorrectly, go to problem 54.

54. A square has a side 20 cm long. What is its perimeter?

Answer: 80 cm

If you answered correctly, go to problem 55.
If you answered incorrectly, review "Finding the Perimeter" in "Squares and Rectangles" on page 105.

55. A rectangle is 18 inches long and 8 inches wide. What is its area?

Answer: 144 in^2

If you answered correctly, go to problem 57.
If you answered incorrectly, go to problem 56.

56. A square has one side 9 cm long. What is its area?

Answer: 81 cm^2

If you answered correctly, go to problem 57.
If you answered incorrectly, review "Finding the Area" in "Squares and Rectangles" on page 106.

57. $\triangle DEF$ is a right triangle with legs 10 and 24 cm long and a hypotenuse of 26 cm. Find the perimeter of $\triangle DEF$.

Answer: 60 cm

If you answered correctly, go to problem 59.
If you answered incorrectly, go to problem 58.

58. $\triangle QRS$ has sides 11, 13, and 17 cm long. Find its perimeter.

Answer: 41 cm

If you answered correctly, go to problem 59.
If you answered incorrectly, review "Finding the Perimeter" in "Triangles" on page 109.

59. $\triangle QRS$ has sides 11, 13, and 17 cm long. It's altitude to the 13 cm side is 10 cm, Find its area.

Answer: 65 cm^2

If you answered correctly, go to problem 61.
If you answered incorrectly, go to problem 60.

60. Find the area of right $\triangle DEF$ with legs 10 and 24 cm long and a hypotenuse of 26 cm.

Answer: 120 cm^2

If you answered correctly, go to problem 61.
If you answered incorrectly, review "Finding the Area" in "Triangles" on page 109.

61. $\square ABCD$ has an obtuse angle at *C*. CD = 17 cm, BC = 12 cm. Height CX is 6 cm. Find its area.

Answer: 102 cm^2

If you answered correctly, go to problem 63.
If you answered incorrectly, go to problem 62.

62. $\square ABCD$ has an obtuse angle at *D*. CD = 20 in., *AD* = 8 in. Height *DX* is 5 in. Find its area.

Answer: 100 in^2

If you answered correctly, go to problem 63.
If you answered incorrectly, review "Finding the Area" in "Parallelograms" on page 112.

63. A trapezoid has sides 6 m, 12 m, 8 m, 8 m, and altitude 6 m. Find its area.

Answer: 54 m^2

If you answered correctly, go to problem 65.
If you answered incorrectly, go to problem 64.

64. A trapezoid has sides 9 ft., 13 ft., 11 ft., 11 ft., and altitude 8 ft. Find its area.

Answer: 88 ft^2

If you answered correctly, go to problem 65.
If you answered incorrectly, review "Finding the Area" in "Trapezoids" on page 115.

65. A regular hexagon has sides 12 inches long, a radius 10 inches long, and an apothem 8 inches long. What is its area?

Answer: 288 in^2

If you answered correctly, go to problem 67.
If you answered incorrectly, go to problem 66.

66. A regular octagon has sides 10 cm long and an apothem 8 cm long. What is its area?

Answer: 320 cm^2

If you answered correctly, go to problem 67.
If you answered incorrectly, review "Finding the Area" in "Regular Polygons" on page 117.

67. Find the circumference of a circle with radius 10.

Answer: 20π

If you answered correctly, go to problem 69.
If you answered incorrectly, go to problem 68.

68. Find the circumference of a circle with diameter 15.

Answer: 15π

If you answered correctly, go to problem 69.
If you answered incorrectly, review "Finding Circumference" on page 119.

69. Find the area of a circle with diameter 14 cm.

Answer: 49π cm^2

If you answered correctly, go to problem 71.
If you answered incorrectly, go to problem 70.

70. Find the area of a circle with radius 10 mm.

Answer: 100π mm^2

If you answered correctly, go to problem 71.
If you answered incorrectly, review "Finding the Area" in "Circles" on page 120.

71. A bus contains 18 boys and 24 girls. What is the ratio of girls to boys?

Answer: 4:3 or $\frac{4}{3}$

If you answered correctly, go to problem 73.
If you answered incorrectly, go to problem 72.

72. A 200 foot length of rope is divided into three parts in the ratio 2:3:5. How long is the shortest length of rope?

Answer: 40 ft.

If you answered correctly, go to problem 73.
If you answered incorrectly, review "Ratio" on page 123.

73. Find the value of x in the proportion $\frac{x}{7} = \frac{35}{49}$.

Answer: 5

If you answered correctly, go to problem 75.
If you answered incorrectly, go to problem 74.

74. If $\frac{5}{x} = \frac{8}{y}$, find the value of $\frac{x}{y}$.

Answer: $\frac{5}{8}$

If you answered correctly, go to problem 75.
If you answered incorrectly, review "Properties of Proportions" on page 125.

75. Rectangle *ABCD* is 4 inches long and 5 inches wide. Rectangle *FGHI* is 8 inches long and 9 inches wide. Are the two rectangles similar?

Answer: no

If you answered correctly, go to problem 77.
If you answered incorrectly, go to problem 76.

76. Regular pentagon *ABCDE* has sides 12 cm long. Regular pentagon *LMNOP* has sides 15 cm long. Are the two pentagons similar?

Answer: yes

If you answered correctly, go to problem 77.
If you answered incorrectly, review "Similar Polygons" on page 128.

77. What must you find to prove two triangles similar?

Answer: Two angles of one equal two angles of another.

If you answered correctly, go to problem 79.
If you answered incorrectly, go to problem 78.

78. Two triangles are formed by a pair of lines intersecting between two parallel lines and the parallel lines themselves. Must the triangles be similar?

Answer: yes

If you answered correctly, go to problem 79.
If you answered incorrectly, review "Similar Triangles" on page 129.

79. Suppose that $CE = 9$ cm, $BE = 5$ cm, and $CD = 10$ cm. Find the length of AD.

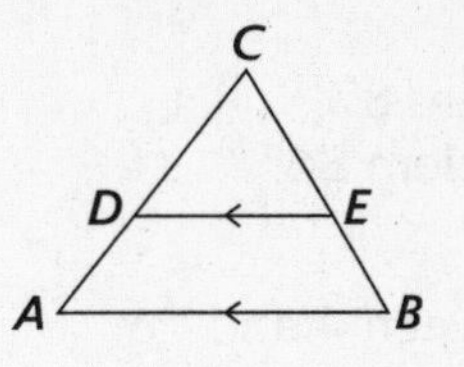

Answer: $\frac{50}{9}$ cm or $5\frac{5}{9}$ cm

If you answered correctly, go to problem 81.
If you answered incorrectly, go to problem 80.

80. Using the same figure that you used for 79, suppose that CE = 12 cm, BE = 4 cm, and CD = 9 cm. Find the length of AC.

Answer: 12 cm

If you answered correctly, go to problem 81.
If you answered incorrectly, review "Proportional Parts of Triangles" on page 132.

81. $\triangle QRS \sim \triangle MNO$. The shortest side of $\triangle QRS$ is 15 mm long, and its longest side is 25. The shortest side of $\triangle MNO$ is 21 mm long. The altitude of the larger triangle is 14 mm long. What is the length of the corresponding altitude of $\triangle QRS$?

Answer: 10 mm

If you answered correctly, go to problem 83.
If you answered incorrectly, go to problem 82.

82. $\triangle LMN \sim \triangle TUV$. The median from $\angle L$ to $\overline{MN}$ is 8 cm long. The median from $\angle T$ to $\overline{UV}$ is 40 cm long. Side LM of the first triangle is 9 cm long. Its corresponding side in the second triangle is TU. How long is TU?

Answer: 45 cm

If you answered correctly, go to problem 83.
If you answered incorrectly, review "Proportional Parts of Similar Triangles" on page 137.

83. $\triangle XYZ \sim \triangle MNO$. An altitude of the first is 12 in., and the corresponding altitude of the second is 20 in. What is the ratio of the areas of $\triangle XYZ$ to $\triangle MNO$?

Answer: 9:25

If you answered correctly, go to problem 85.
If you answered incorrectly, go to problem 84.

84. $\triangle ABC \sim \triangle GHI$. An altitude of the first is 24 in., and the altitude of the second is 18 in. What is the ratio of the areas of $\triangle GHI$ to $\triangle ABC$?

Answer: 9:16

If you answered correctly, go to problem 85.
If you answered incorrectly, review "Perimeter and Area of Similar Triangles" on page 139.

85. 12 is the geometric mean between 6 and what other number?

Answer: 24

If you answered correctly, go to problem 87.
If you answered incorrectly, go to problem 86.

86. Find the mean proportional between 4 and 16.

Answer: 8

If you answered correctly, go to problem 87.
If you answered incorrectly, review "Geometric Mean" on page 143.

Use the following figure to solve 87 and 88 (if necessary).

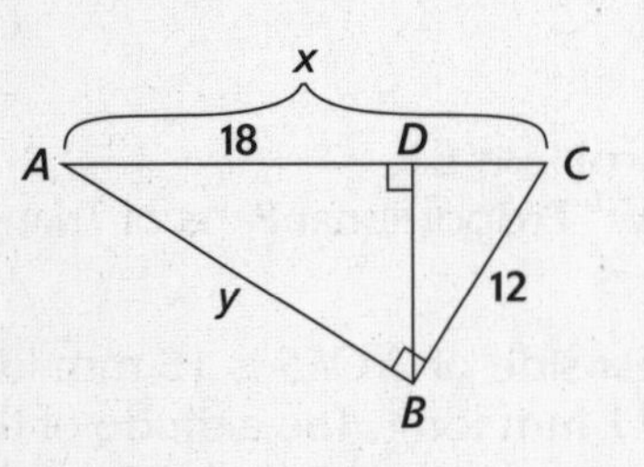

87. Find the length of x.

Answer: 24

If you answered correctly, go to problem 89.
If you answered incorrectly, go to problem 88.

88. Find the length of y.

Answer: $12\sqrt{3}$

If you answered correctly, go to problem 89.
If you answered incorrectly, review "Altitude to the Hypotenuse" on page 144.

89. A right triangle has one leg 15 in. and one 36 in. How long is the hypotenuse?

Answer: 39 in.

If you answered correctly, go to problem 91.
If you answered incorrectly, go to problem 90.

90. A right triangle has one leg 15 in. long and a hypotenuse 25 in. long. How long is the other leg?

Answer: 20 in.

If you answered correctly, go to problem 91.
If you answered incorrectly, review "The Pythagorean Theorem" on page 147.

91. A right triangle has legs 9 cm long and 12 cm long. How long is its hypotenuse?

Answer: 15 cm

If you answered correctly, go to problem 93.
If you answered incorrectly, go to problem 92.

92. An isosceles right triangle has one leg that is 6 inches long. How long are the other two sides?

Answer: 6 inches, $6\sqrt{2}$ in.

If you answered correctly, go to problem 93.
If you answered incorrectly, review "Isosceles Right Triangle" on page 153.

93. The hypotenuse of a right isosceles triangle is 10 cm. What is the length of each leg?

Answer: $5\sqrt{2}$ cm

If you answered correctly, go to problem 95.
If you answered incorrectly, go to problem 94.

94. The hypotenuse of a right isosceles triangle is 24 cm. What is the length of each leg?

Answer: $12\sqrt{2}$ cm

If you answered correctly, go to problem 95.
If you answered incorrectly, review "Isosceles Right Triangle" on page 153.

95. A 30-60-90 triangle's shortest side is 13 inches long. What is the length of the triangle's second shortest side?

Answer: $13\sqrt{3}$ in.

If you answered correctly, go to problem 97.
If you answered incorrectly, go to problem 96.

96. Find the altitude of an equilateral triangle with a perimeter of 24 cm.

Answer: $4\sqrt{3}$ cm

If you answered correctly, go to problem 97.
If you answered incorrectly, review "30-60-90 Right Triangle" on page 155.

97. What is the name given to a chord that extends beyond a circle?

Answer: secant

If you answered correctly, go to problem 99.
If you answered incorrectly, go to problem 98.

98. What is the name of the longest chord in a circle?

Answer: diameter

If you answered correctly, go to problem 99.
If you answered incorrectly, review "Parts of a Circle" on page 159.

99. What makes a common tangent to two circles an internal common tangent?

Answer: It crosses the line connecting the circles' centers.

If you answered correctly, go to problem 101.
If you answered incorrectly, go to problem 100.

100. What is true of a radius to a point of tangency and the tangent to the circle?

Answer: They form right angles.

If you answered correctly, go to problem 101.
If you answered incorrectly, review "Parts of a Circle" on page 159.

101. What is the length of a 90° central angle's arc in a circle of radius 12?

Answer: 6π

If you answered correctly, go to problem 103.
If you answered incorrectly, go to problem 102.

102. What is the degree measure of a 120° central angle's arc in a circle of radius 12?

Answer: 120°

If you answered correctly, go to problem 103.
If you answered incorrectly, review "Central Angles and Arcs" on page 161.

103. $\angle ABC$ is an inscribed angle with an intercepted arc of 120°. What is its degree measure?

Answer: 60°

If you answered correctly, go to problem 105.
If you answered incorrectly, go to problem 104.

104. What kind of inscribed angle intercepts a semicircle?

Answer: a right angle, or an angle of 90°

If you answered correctly, go to problem 105.
If you answered incorrectly, review "Arcs and Inscribed Angles" on page 164.

Use the following figure to answer 105 and, if necessary, 106.

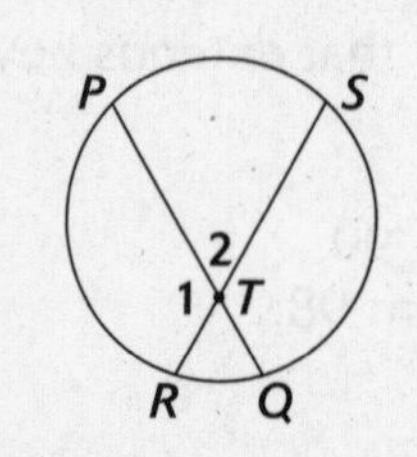

105. $m\overset{\frown}{QR} = 45°$ and $m\overset{\frown}{PS} = 65°$. Find $m\angle 2$.

Answer: 55°

If you answered correctly, go to problem 107.
If you answered incorrectly, go to problem 106.

106. $m\widehat{PR} = 100°$ and $m\widehat{QS} = 120°$. Find $m\angle 1$.

Answer: 110°

If you answered correctly, go to problem 107.
If you answered incorrectly, review "Angles Formed by Chords, Secants, and Tangents" on page 167.

Use the following figure to answer 107 and, if necessary, 108.

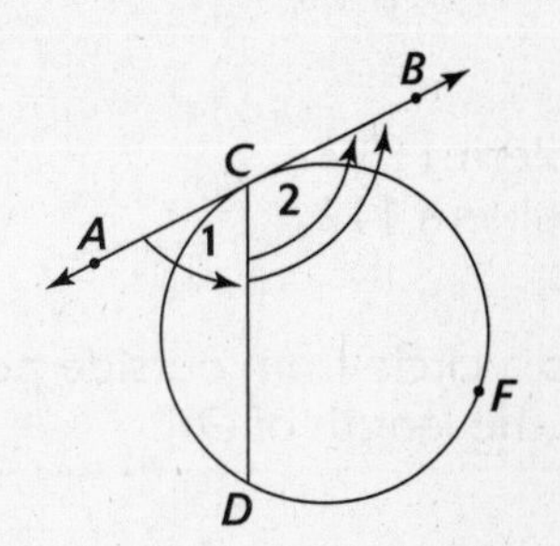

107. $m\widehat{CD} = 150°$ and $m\widehat{CFD} = 210°$. Find $m\angle 1$.

Answer: 75°

If you answered correctly, go to problem 109.
If you answered incorrectly, go to problem 108.

108. $m\widehat{CD} = 120°$ and $m\widehat{CFD} = 240°$. Find $m\angle 2$.

Answer: 120°

If you answered correctly, go to problem 109.
If you answered incorrectly, review "Angles Formed by Chords, Secants, and Tangents" on page 167.

Use the following figure to answer 109 and, if necessary, 110.

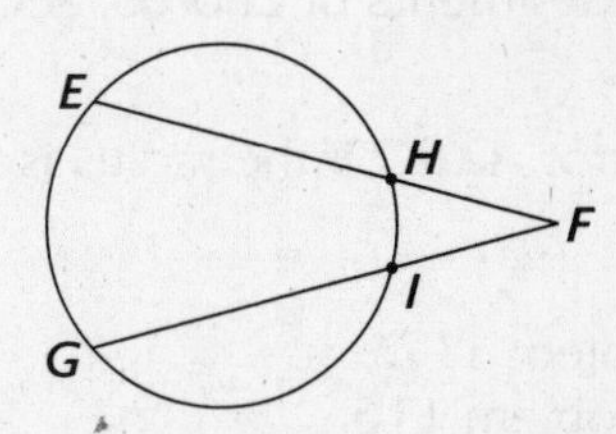

109. $m\widehat{EG} = 85°$ and $m\widehat{HI} = 25°$. Find $m\angle EFG$.

Answer: 30°

If you answered correctly, go to problem 111.
If you answered incorrectly, go to problem 110.

110. $m\widehat{EG} = 75°$ and $m\widehat{HI} = 40°$. Find $m\angle EFG$.

Answer: 17.5°

If you answered correctly, go to problem 111.
If you answered incorrectly, review "Angles Formed by Chords, Secants, and Tangents" on page 167.

111. *LM* and *NM* are secants to circle *O*. *LP* and *NR* are chords and are part of *LM* and *NM* respectively. $LM = 36$, $MP = 8$, $MR = 6$. Find *MN*.

Answer: 48

If you answered correctly, go to problem 113.
If you answered incorrectly, go to problem 112.

112. *QS* and *QT* are two tangents to a circle from outside point *Q*. They intercept an arc of 150°. If *QS* is 21 cm long, find the length of *QT*.

Answer: 21 cm

If you answered correctly, go to problem 113.
If you answered incorrectly, review "Segments of Chords, Secants, and Tangents" on page 174.

113. A 16 inch long secant intersects with a tangent at a point outside the circle. 12 inches of the secant are inside the circle. How long is the tangent?

Answer: 8 in.

If you answered correctly, go to problem 115.
If you answered incorrectly, go to problem 114.

114. A 24 cm long secant intersects with a tangent at a point outside the circle. 18 cm of the secant are inside the circle. How long is the tangent?

Answer: 12 cm

If you answered correctly, go to problem 115.
If you answered incorrectly, review "Segments of Chords, Secants, and Tangents" on page 174.

115. In circle *O*, find the area of sector *AXBO* if the radius is 9 cm and the central angle is 120°.

Answer: 27π cm^2

If you answered correctly, go to problem 117.
If you answered incorrectly, go to problem 116.

116. In circle *Q*, find the area of sector *CDVQ* if the radius is 5 cm and the central angle is 72°.

Answer: 5π cm^2

If you answered correctly, go to problem 117.
If you answered incorrectly, review "Arc Lengths and Sectors" on page 177.

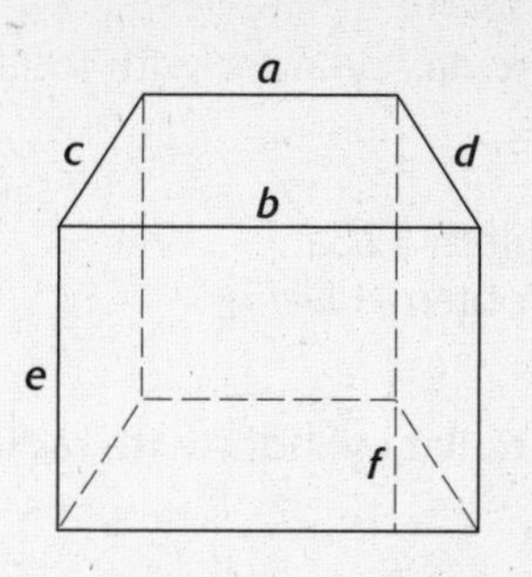

117. Find the total area of the right rectangular prism shown in the preceding figure: $a = 6$ cm, $c = 5$ cm, $e = 7$ cm.

Answer: 214 cm^2

If you answered correctly, go to problem 119.
If you answered incorrectly, go to problem 118.

118. Find the lateral area of the regular right hexagonal prism with height 10 inches and base edges of 6 inches.

Answer: 360 in^2

If you answered correctly, go to problem 119.
If you answered incorrectly, review "Oblique vs. Right Prisms" on page 183.

119. A 12 foot high right rectangular prism has a base that is 4 feet by 6 feet. What is the prism's volume?

Answer: 288 ft^3

If you answered correctly, go to problem 121.
If you answered incorrectly, go to problem 120.

120. An 8 yard high right triangular prism has a base that is a right triangle, 3 yards by 4 yards by 5 yards. What is the prism's volume?

Answer: 48 yd^3

If you answered correctly, go to problem 121.
If you answered incorrectly, review "Volume of a Prism" on page 188.

121. Find the lateral area of the right circular cylinder with height of 6 m and radius 10 m.

Answer: 120π m^2

If you answered correctly, go to problem 123.
If you answered incorrectly, go to problem 122.

122. Find the total area of the right circular cylinder in problem 121.

Answer: 320π m^2

If you answered correctly, go to problem 123.
If you answered incorrectly, review "Right Circular Cylinders" on page 189.

123. Find the volume of the right circular cylinder with radius 5 cm and height 8 cm.

Answer: 200π cm^3

If you answered correctly, go to problem 125.
If you answered incorrectly, go to problem 124.

124. Find the volume of the right circular cylinder with radius 6 cm and height 9 cm.

Answer: 324π cm^3

If you answered correctly, go to problem 125.
If you answered incorrectly, review "Right Circular Cylinders" on page 189.

Problems 125 through 128 refer to the following figure.

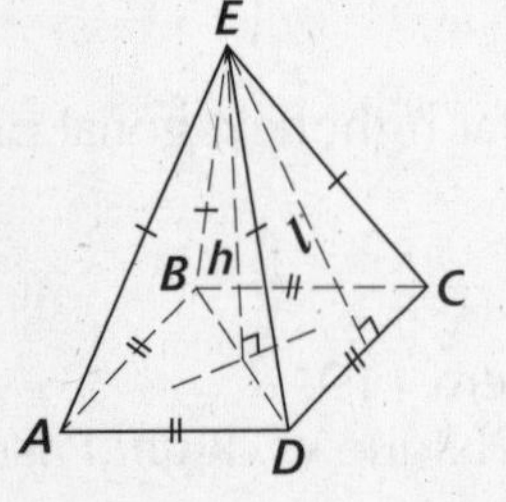

125. Find the lateral area of the regular square pyramid with $AD = 12$, $AE = 10$, $h = 7$, $l = 8$ cm.

Answer: 192 cm^2

If you answered correctly, go to problem 127.
If you answered incorrectly, go to problem 126.

126. Find the total area of the regular square pyramid with $AD = 12$, $AE = 10$, $h = 7$, $l = 8$ cm.

Answer: 336 cm^2

If you answered correctly, go to problem 127.
If you answered incorrectly, review "Pyramids" on page 193.

127. Find the volume of the regular square pyramid with $AD = 10$, $AE = 12$, $h = 9$, $l = 10$ cm.

Answer: 300 cm^3

If you answered correctly, go to problem 129.
If you answered incorrectly, go to problem 128.

128. Find the volume of the regular square pyramid with $AD = 9$, $AE = 12$, $h = 7$, $l = 8$ cm.

Answer: 189 cm^3

If you answered correctly, go to problem 129.
If you answered incorrectly, review "Pyramids" on page 193.

Problems 129 through 132 refer to the following figure.

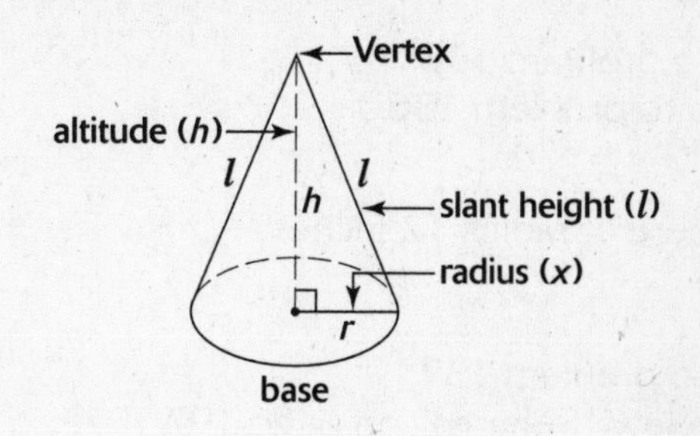

129. Find the lateral area of the right circular cone with $r = 6$, $h = 8$, $l = 10$ inches.

Answer: 60π in^2

If you answered correctly, go to problem 131.
If you answered incorrectly, go to problem 130.

130. Find the total area of the right circular cone with $r = 6$, $h = 8$, $l = 10$ inches.

Answer: 96π in^2

If you answered correctly, go to problem 131.
If you answered incorrectly, review "Right Circular Cones" on page 195.

131. Find the volume of the right circular cone with $r = 6$, $h = 8$, $l = 10$ inches.

Answer: 96π in^3

If you answered correctly, go to problem 133.
If you answered incorrectly, go to problem 132.

132. Find the volume of the right circular cone with $r = 9$, $h = 11$, $l = 12$ feet.

Answer: 297π ft^3

If you answered correctly, go to problem 133.
If you answered incorrectly, review "Right Circular Cones" on page 195.

133. Find the surface area of a sphere of radius 5 inches.

Answer: 100π in^2

If you answered correctly, go to problem 135.
If you answered incorrectly, go to problem 134.

134. Find the surface area of a sphere of radius 8 centimeters.

Answer: 256π cm^2

If you answered correctly, go to problem 135.
If you answered incorrectly, review "Spheres" on page 197.

135. Find the volume of a sphere of radius 9 millimeters.

Answer: 972π mm^3

If you answered correctly, go to problem 137.
If you answered incorrectly, go to problem 136.

136. Find the volume of a sphere of radius 12 inches.

Answer: 2304π in^3

If you answered correctly, go to problem 137.
If you answered incorrectly, review "Spheres" on page 197.

Problems 137 and 138 refer to the following figure.

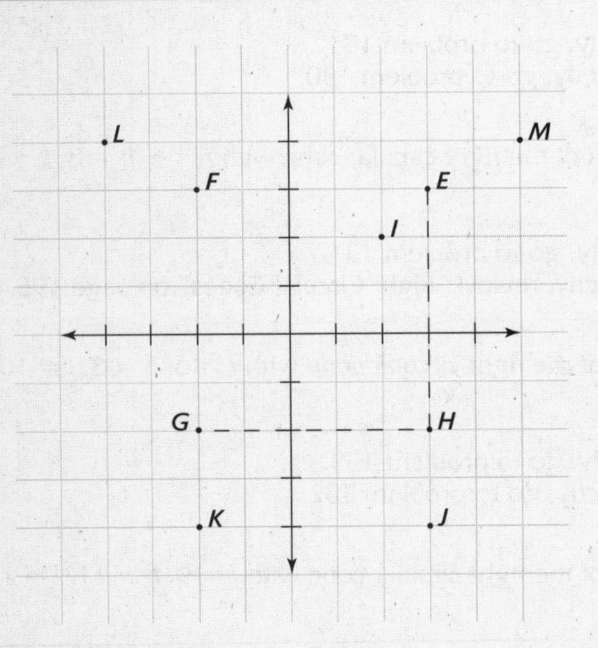

137. What are the coordinates of point L?

Answer: $(-4, 4)$

If you answered correctly, go to problem 139.
If you answered incorrectly, go to problem 138.

138. What are the coordinates of point K?

Answer: $(-2, -4)$

If you answered correctly, go to problem 139.
If you answered incorrectly, review "Locating Points on Coordinate Axes" on page 201.

139. Use the distance formula to find the distance between points with coordinates (6, 4) and (−7, −3).

Answer: $\sqrt{218}$

If you answered correctly, go to problem 141.
If you answered incorrectly, go to problem 140.

140. Use the distance formula to find the distance between points with coordinates (8, 5) and (−4, −4).

Answer: 15

If you answered correctly, go to problem 141.
If you answered incorrectly, review "The Distance Formula" on page 203.

141. Segment *EF* has endpoints *E*(−3, 3) and *F*(7, −5). Find the coordinates of its midpoint, *M*.

Answer: (2, −1)

If you answered correctly, go to problem 143.
If you answered incorrectly, go to problem 142.

142. Segment *GH* has endpoints *G*(−9, 4) and *H*(11, −8). Find the coordinates of its midpoint, *N*.

Answer: (1, −2)

If you answered correctly, go to problem 143.
If you answered incorrectly, review "The Midpoint Formula" on page 205.

143. Points (3, 3) and (0, −5) are on line *a*. Find line *a*'s slope.

Answer: $\frac{8}{3}$

If you answered correctly, go to problem 145.
If you answered incorrectly, go to problem 144.

144. Points (−6, −8) and (4, 2) are on line *l*. Find line *l*'s slope.

Answer: 1

If you answered correctly, go to problem 145.
If you answered incorrectly, review "Slope of a Line" on page 208.

145. A line has a slope of $\frac{-4}{3}$. For a line to be perpendicular to it, that line must have a slope of _____?

Answer: $\frac{3}{4}$

If you answered correctly, go to problem 147.
If you answered incorrectly, go to problem 146.

146. A line has a slope of $\frac{1}{4}$. For a line to be perpendicular to it, that line must have a slope of _____?

Answer: −4

If you answered correctly, go to problem 147.
If you answered incorrectly, review "Slopes of Parallel and Perpendicular Lines" on page 210.

147. Find the *x*-intercept of $ax + by = c$.

Answer: $x = \frac{c}{a}$

If you answered correctly, go to problem 149.
If you answered incorrectly, go to problem 148.

148. Find the y-intercept of $ax + by = c$.

Answer: $y = \frac{c}{b}$

If you answered correctly, go to problem 149.
If you answered incorrectly, review "Standard Form of a Linear Equation" on page 212.

149. Write the point-slope form of a linear equation.

Answer: $y - y_1 = m(x - x_1)$

If you answered correctly, go to problem 151.
If you answered incorrectly, go to problem 150.

150. Write the slope-y-intercept form of a linear equation.

Answer: $y = mx + b$

If you answered correctly, go to problem 151.
If you answered incorrectly, review "Equations of Lines" on page 211.

151. Find the equation of a line containing the points (3, 4) and (−1, −4) and write it in the point-slope form.

Answer: $y - 4 = 2(x - 3)$ or $y + 4 = 2(x + 1)$

If you answered correctly, you are finished! Congratulations.
If you answered incorrectly, go to problem 152.

152. Rewrite the equation from 151 in slope-y-intercept form.

Answer: $y = 2x - 2$

If you answered correctly, you are finished! Congratulations.
If you answered incorrectly, review "Equations of Lines" on page 211.

Appendix

Postulates and Theorems

Chapter 1

Postulate 1: A line contains at least two points.

Postulate 2: A plane contains a minimum of three noncollinear points.

Postulate 3: Through any two points there can be exactly one line.

Postulate 4: Through any three noncollinear points there can be exactly one plane.

Postulate 5: If two points lie in a plane, then the line they lie on is in the same plane.

Postulate 6: Where two planes intersect, their intersection is a line.

Postulate 7 (The Ruler Postulate): Each point on a line or a segment can be paired with a single real number, which is known as that point's coordinate. The distance between two points is the absolute value of the difference of their coordinates.

Postulate 8 (The Segment Addition Postulate):

If N lies between M and P on a line, then $MN + NP = MP$.

Postulate 9 (The Protractor Postulate): Supposes that a point, Z, exists on line XY. Each ray with endpoint Z that exists on one side of line XY may be paired with exactly one number between 0° and 180°. The positive difference between two numbers representing two different rays is the degree measure of the angle with those rays as its sides.

Postulate 10 (Addition of Angles Postulate):

If $\overrightarrow{ZS}$ lies between $\overrightarrow{ZX}$ and $\overrightarrow{ZT}$, then $m\angle XZT = m\angle XZS + m\angle SZT$.

Theorem 1: If two lines intersect, they intersect in exactly one point.

Theorem 2: If a point lies outside a line, then exactly one plane contains the line and the point.

Theorem 3: If two lines intersect, then exactly one plane contains both lines.

Theorem 4: The midpoint of a line segment is the point that's an equal distance from both endpoints.

Theorem 5: An angle that is not a straight angle has only one angle bisector.

Theorem 6: All right angles are equal.

Theorem 7: Vertical angles are equal in measure.

Theorem 8: If two angles are complements of the same or equal angles, they are equal to each other.

Theorem 9: If two adjacent angles have their noncommon sides lying on a line, then they are supplementary angles.

Theorem 10: If two angles are supplements of the same or equal angles, they are equal to each other.

Theorem 11: If two lines are parallel to a third line, they are parallel to each other.

Chapter 2

Postulate 11 (The Parallel Postulate): If two parallel lines are cut by a transversal, their corresponding angles are equal.

Postulate 12: If two lines and a transversal form equal corresponding angles, then the lines are parallel.

Theorem 12: If parallel lines are cut by a transversal, their alternate interior angles are equal.

Theorem 13: If parallel lines are cut by a transversal, their alternate exterior angles are equal.

Theorem 14: If parallel lines are cut by a transversal, their consecutive interior angles are supplementary.

Theorem 15: If parallel lines are cut by a transversal, their consecutive exterior angles are supplementary.

Theorem 16: If two lines and a transversal form equal alternate interior angles, then the lines are parallel.

Theorem 17: If two lines and a transversal form equal alternate exterior angles, then the lines are parallel.

Theorem 18: If two lines and a transversal form consecutive interior angles that are supplementary, then the lines are parallel.

Theorem 19: If two lines and a transversal form consecutive exterior angles that are supplementary, then the lines are parallel.

Theorem 20: In a plane, if two lines are perpendicular to the same line, then the lines are parallel.

Chapter 3

Postulate 13 (SSS): If each side of one triangle is congruent to the corresponding side of a second triangle, then the two triangles are congruent.

Postulate 14 (SAS): If two sides of one triangle and the angle between them are congruent to the corresponding parts of a second triangle, then the two triangles are congruent.

Postulate 15 (ASA): If two angles of one triangle and the side between them are congruent to the corresponding parts of a second triangle, then the two triangles are congruent.

Postulate 16 (SAA): If two angles of one triangle and a side not between them are congruent to the corresponding parts of a second triangle, then the two triangles are congruent.

Postulate 17 (HL): If the hypotenuse and one leg of a right triangle are congruent to the corresponding parts of a second right triangle, then the two triangles are congruent.

Theorem 21: The sum of the interior angles of any triangle is 180°.

Theorem 22: An exterior angle of a triangle is equal to the sum of the remote interior angles.

Theorem 23: Each angle of an equiangular triangle has a measure of 60°.

Theorem 24 (HA): If the hypotenuse and an acute angle of one right triangle are congruent to the corresponding parts of a second right triangle, then the two triangles are congruent.

Theorem 25 (LL): If the legs of one right triangle are congruent to the corresponding parts of a second right triangle, then the two triangles are congruent.

Theorem 26 (LA): If one leg and an acute angle of one right triangle are congruent to the corresponding parts of a second right triangle, then the two triangles are congruent.

Theorem 27: If two sides of a triangle are equal, then the angles opposite those sides are equal.

Theorem 28: If two angles of a triangle are equal, then the sides opposite those angles are equal.

Theorem 29: If a triangle is equilateral, then it is also equiangular.

Theorem 30: If a triangle is equiangular, then it is also equilateral.

Theorem 31: If two sides of a triangle are unequal, then the measures of the angles opposite those sides are unequal, and the greater angle is opposite the greater side.

Theorem 32: If two angles of a triangle are unequal, then the measures of the sides opposite those angles are unequal, and the longer side is opposite the greater angle.

Theorem 33: The sum of the lengths of any two sides of a triangle is greater than the length of the third side.

Chapter 4

No postulates.

Theorem 34: If a convex polygon has n sides, then its interior angle sum is given by the equation:

$$S = (n - 2) \times 180°$$

Theorem 35: If a polygon is convex, then the sum of the exterior angles (drawn one per vertex) is 360°.

Theorem 36: Either diagonal of a parallelogram divides it into two congruent triangles.

Theorem 37: Opposite sides of a parallelogram are congruent.

Theorem 38: Opposite angles of a parallelogram are congruent.

Theorem 39: Consecutive angles of a parallelogram are supplementary.

Theorem 40: The diagonals of a parallelogram bisect each other.

Theorem 41: If both pairs of opposite sides of a quadrilateral are equal, then it is a parallelogram.

Theorem 42: If both pairs of opposite angles of a quadrilateral are equal, then it is a parallelogram.

Theorem 43: If all pairs of consecutive angles of a quadrilateral are supplementary, then it is a parallelogram.

Theorem 44: If one pair of opposite sides of a quadrilateral are both equal and parallel, then it is a parallelogram.

Theorem 45: If the diagonals of a quadrilateral bisect each other, then it is a parallelogram.

Theorem 46: The diagonals of a rectangle are equal.

Theorem 47: The diagonals of a rhombus bisect the opposite angles.

Theorem 48: The diagonals of a rhombus are perpendicular to one another.

Theorem 49: Base angles of an isosceles trapezoid are equal.

Theorem 50: Diagonals of an isosceles trapezoid are equal.

Theorem 51: The median of any trapezoid is parallel to both bases.

Theorem 52: The median of any trapezoid is half the length of the sum of the bases.

Theorem 53 (The Midpoint Theorem): The segment joining the midpoints of any two sides of a triangle is half the length of the third side and parallel to it.

Chapter 5

No postulates or theorems.

Chapter 6

Postulate 18 (The AA Similarity Postulate): If two angles of one triangle are equal to two angles of a second triangle, then the triangles are similar.

Theorem 54 (The Side Splitting Theorem): If a line (or segment) is parallel to one side of a triangle and intersects the other two sides, it divides those sides proportionally.

Theorem 55 (Angle Bisector Theorem): If a ray bisects an angle of a triangle, then it divides the opposite side into segments that are proportional to the sides forming the triangle.

Theorem 56: If two triangles are similar, then the ratio of any two corresponding segments in those triangles (altitudes, medians, and angle bisectors) are equal to the ratio between any two corresponding sides.

Theorem 57: If two similar triangles have a scale factor of $a{:}b$, then the ratio of their perimeters is $a{:}b$.

Theorem 58: If two similar triangles have a scale factor of $a{:}b$, then the ratio of their areas is $a^2{:}b^2$.

Chapter 7

No postulates.

Theorem 59: If the altitude is drawn to the hypotenuse of a right triangle, it creates two similar right triangles, each similar to the original triangle as well as to each other.

Theorem 60: If the altitude is drawn to the hypotenuse of a right triangle, then each leg is the geometric mean between the hypotenuse and the segment on the hypotenuse that it touches.

Theorem 61: If an altitude is drawn to the hypotenuse of a right triangle, then it is the geometric mean between the segments of the hypotenuse.

Theorem 62 (The Pythagorean Theorem): The square *on* the hypotenuse of a right triangle *is equal in area* to the *sum of the areas* of the squares on the two legs.

Theorem 63: If a triangle has sides of lengths a, b, and c where c is the longest length, and $c^2 = a^2 + b^2$, then that triangle is a right triangle, and c is its hypotenuse.

Theorem 64: If a triangle has sides of lengths a, b, and c where c is the longest length, and $c^2 > a^2 + b^2$, then that triangle is an obtuse triangle.

Theorem 65: If a triangle has sides of lengths a, b, and c where c is the longest length, and $c^2 < a^2 + b^2$, then that triangle is an acute triangle.

Chapter 8

Postulate 19 (Arc Addition Postulate): If B is a point on $\overparen{ABC}$, then $\overparen{AB} + \overparen{BC} = \overparen{ABC}$.

Theorem 66: In the same or equal circles, if two central angles have equal measures, then their corresponding minor arcs have equal measures.

Theorem 67: In the same or equal circles, if two minor arcs have equal measures, then their corresponding central angles have equal measure.

Theorem 68: In a circle, the degree measure of an inscribed angle is half the degree measure of its intercepted arc.

Theorem 69: If two inscribed angles intercept the same or equal arcs, then the angles are of equal degree measure.

Theorem 70: If an inscribed angle intercepts a semicircle, then its measure is 90°.

Theorem 71: If a tangent and a diameter (or radius) of a circle meet at the point of tangency, then they are perpendicular to each other.

Theorem 72: If a chord and a tangent to a circle are perpendicular at the point of tangency, then the chord is a diameter of that circle.

Theorem 73: The measure of an angle formed by two chords intersecting inside a circle is equal to half the sum of the degree measures of the intercepted arcs of the angle and its vertical angle.

Theorem 74: The degree measure of an angle formed by a tangent and a chord meeting at the point of tangency is half the degree measure of the intercepted arc.

Theorem 75: When two secants intersect outside a circle, the degree measure of the angle they form is one-half the difference of the degree measure of their intercepted arcs.

Theorem 76: In the same or equal circles, if two chords are equal in measure, then the minor arcs with the same endpoints are equal in measure.

Theorem 77: In the same or equal circles, if two minor arcs are equal in measure, then the chords with the same endpoints are equal in measure.

Theorem 78: If a diameter is perpendicular to a chord, then it bisects the chord and its arcs.

Theorem 79: In the same or equal circles, if two chords are equal in measure, then they are equidistant from the center.

Theorem 80: In the same or equal circles, if two chords are equidistant from the center, they are equal in measure.

Theorem 81: If two chords intersect inside a circle, then the product of the segments of one chord equals the product of the segments of the other chord.

Theorem 82: If two secants to a circle intersect at an outside point, the product of one secant and its external portion is equal to the product of the other secant and its external portion.

Theorem 83: If a tangent and a secant to the same circle intersect at an outside point, then the square of the measure of the tangent equals the product of the measure of the secant and its external portion.

Theorem 84: If two tangents to the same circle intersect at an outside point, then they are equal in measure.

Chapter 9

No postulates.

Theorem 85: The lateral area, LA, of a right prism of altitude h and perimeter p is given by the equation:

$$LA = ph \text{ units}^2$$

Theorem 86: The total area, TA, of a right prism with lateral area LA and a base area B is given by the equation:

$$TA = LA + 2B \text{ or } TA = ph + 2B$$

Theorem 87: The volume of a right prism with a base area B and an altitude h is given by the equation:

$$V = Bh \text{ units}^3$$

Theorem 88: The lateral area, LA, of a right circular cylinder with a base circumference of C and altitude h is given by the equation:

$$LA = Ch \text{ units}^2\text{, or } LA = 2\pi rh \text{ units}^2$$

Theorem 89: The total area, TA, of a right circular cylinder with lateral area, LA, and a base area B is given by the equation:

$$TA = LA + 2B \text{ units}^2 \text{ or}$$
$$TA = 2\pi rh + 2\pi r^2 \text{ units}^2 \text{ or}$$
$$TA = 2\pi r(h + r) \text{ units}^2$$

Theorem 90: The volume of a right circular cylinder, V, with base area B and altitude h is given by the equation:

$$V = Bh \text{ units}^3 \text{ or}$$
$$V = \pi r^2 h \text{ units}^3$$

Theorem 91: The lateral area, LA, of a regular pyramid with slant height l and base perimeter p is given by the equation:

$$LA = \frac{1}{2} pl \text{ units}^2$$

Theorem 92: The total area, TA, of a regular pyramid with lateral area LA and a base area B is given by the equation:

$$TA = LA + B \text{ units}^2 \text{ or}$$
$$TA = \frac{1}{2} pl + B \text{ units}^2$$

Theorem 93: The volume of a regular pyramid with a base area B and an altitude h is given by the equation:

$$V = \frac{1}{3} Bh \text{ units}^3$$

Chapter 10

No postulates.

Theorem 94 (The Distance Formula): If the coordinates of two points are (x_1, y_1) and (x_2, y_2), then the distance, d, between the two points is expressed by the formula:

$$d = \sqrt{(x_2 - x_1)^2 + (y_2 - y_1)^2}$$

Theorem 95 (The Midpoint Formula): If the coordinates of two endpoints of a segment are given by (x_1, y_1) and (x_2, y_2), respectively, then the midpoint of that segment, M, may be found by the formula

$$M = \left(\frac{x_1 + x_2}{2}, \frac{y_1 + y_2}{2}\right)$$

Theorem 96: If two nonvertical lines are parallel, then their slopes are the same.

Theorem 97: If two lines have the same slope, then the lines are parallel.

Theorem 98: If two nonvertical lines are perpendicular, then their slopes are negative reciprocals of one another, that is, the product of the slopes is -1.

Theorem 99: If the slopes of two lines multiply together to form a product of -1, then the lines are perpendicular and nonvertical.

Theorem 100: The point-slope form of a nonvertical line passing through (x, y) and having slope m is

$$y - y_1 = m(x - x_1)$$

Theorem 101: The slope-y-intercept form of a nonvertical line with slope m and y-intercept vertical coordinate b is $y = mx + b$.

Index

D

E

G

H

I

L

M

N

O

P

Q

R

S

T

U

V

X

Y

Wiley Publishing, Inc.
End-User License Agreement

READ THIS. You should carefully read these terms and conditions before opening the software packet(s) included with this book "Book". This is a license agreement "Agreement" between you and Wiley Publishing, Inc. "WPI". By opening the accompanying software packet(s), you acknowledge that you have read and accept the following terms and conditions. If you do not agree and do not want to be bound by such terms and conditions, promptly return the Book and the unopened software packet(s) to the place you obtained them for a full refund.

1. **License Grant.** WPI grants to you (either an individual or entity) a nonexclusive license to use one copy of the enclosed software program(s) (collectively, the "Software") solely for your own personal or business purposes on a single computer (whether a standard computer or a workstation component of a multi-user network). The Software is in use on a computer when it is loaded into temporary memory (RAM) or installed into permanent memory (hard disk, CD-ROM, or other storage device). WPI reserves all rights not expressly granted herein.

2. **Ownership.** WPI is the owner of all right, title, and interest, including copyright, in and to the compilation of the Software recorded on the physical packet included with this Book "Software Media". Copyright to the individual programs recorded on the Software Media is owned by the author or other authorized copyright owner of each program. Ownership of the Software and all proprietary rights relating thereto remain with WPI and its licensers.

3. **Restrictions on Use and Transfer.**

   (a) You may only (i) make one copy of the Software for backup or archival purposes, or (ii) transfer the Software to a single hard disk, provided that you keep the original for backup or archival purposes. You may not (i) rent or lease the Software, (ii) copy or reproduce the Software through a LAN or other network system or through any computer subscriber system or bulletin-board system, or (iii) modify, adapt, or create derivative works based on the Software.

   (b) You may not reverse engineer, decompile, or disassemble the Software. You may transfer the Software and user documentation on a permanent basis, provided that the transferee agrees to accept the terms and conditions of this Agreement and you retain no copies. If the Software is an update or has been updated, any transfer must include the most recent update and all prior versions.

4. **Restrictions on Use of Individual Programs.** You must follow the individual requirements and restrictions detailed for each individual program on the Software Media. These limitations are also contained in the individual license agreements recorded on the Software Media. These limitations may include a requirement that after using the program for a specified period of time, the user must pay a registration fee or discontinue use. By opening the Software packet(s), you agree to abide by the licenses and restrictions for these individual programs that are detailed on the Software Media. None of the material on this Software Media or listed in this Book may ever be redistributed, in original or modified form, for commercial purposes.

5. **Limited Warranty.**

   (a) WPI warrants that the Software and Software Media are free from defects in materials and workmanship under normal use for a period of sixty (60) days from the date of purchase of this Book. If WPI receives notification within the warranty period of defects in materials or workmanship, WPI will replace the defective Software Media.

   (b) WPI AND THE AUTHOR(S) OF THE BOOK DISCLAIM ALL OTHER WARRANTIES, EXPRESS OR IMPLIED, INCLUDING WITHOUT LIMITATION IMPLIED WARRANTIES OF MERCHANTABILITY AND FITNESS FOR A PARTICULAR PURPOSE, WITH RESPECT TO THE SOFTWARE, THE PROGRAMS, THE SOURCE CODE CONTAINED THEREIN, AND/OR THE TECHNIQUES DESCRIBED IN THIS BOOK. WPI DOES NOT WARRANT THAT THE FUNCTIONS CONTAINED IN THE SOFTWARE WILL MEET YOUR REQUIREMENTS OR THAT THE OPERATION OF THE SOFTWARE WILL BE ERROR FREE.

   (c) This limited warranty gives you specific legal rights, and you may have other rights that vary from jurisdiction to jurisdiction.

6. **Remedies.**

   (a) WPI's entire liability and your exclusive remedy for defects in materials and workmanship shall be limited to replacement of the Software Media, which may be returned to WPI with a copy of your receipt at the following address: Software Media Fulfillment Department, Attn.: *CliffsNotes® Geometry Practice Pack,* Wiley Publishing, Inc., 10475 Crosspoint Blvd., Indianapolis, IN 46256, or call 1-877-762-2974. Please allow four to six weeks for delivery. This Limited Warranty is void if failure of the Software Media has resulted from accident, abuse, or misapplication. Any replacement Software Media will be warranted for the remainder of the original warranty period or thirty (30) days, whichever is longer.

   (b) In no event shall WPI or the author be liable for any damages whatsoever (including without limitation damages for loss of business profits, business interruption, loss of business information, or any other pecuniary loss) arising from the use of or inability to use the Book or the Software, even if WPI has been advised of the possibility of such damages.

   (c) Because some jurisdictions do not allow the exclusion or limitation of liability for consequential or incidental damages, the above limitation or exclusion may not apply to you.

7. **U.S. Government Restricted Rights.** Use, duplication, or disclosure of the Software for or on behalf of the United States of America, its agencies and/or instrumentalities "U.S. Government" is subject to restrictions as stated in paragraph (c)(1)(ii) of the Rights in Technical Data and Computer Software clause of DFARS 252.227-7013, or subparagraphs (c) (1) and (2) of the Commercial Computer Software - Restricted Rights clause at FAR 52.227-19, and in similar clauses in the NASA FAR supplement, as applicable.

8. **General.** This Agreement constitutes the entire understanding of the parties and revokes and supersedes all prior agreements, oral or written, between them and may not be modified or amended except in a writing signed by both parties hereto that specifically refers to this Agreement. This Agreement shall take precedence over any other documents that may be in conflict herewith. If any one or more provisions contained in this Agreement are held by any court or tribunal to be invalid, illegal, or otherwise unenforceable, each and every other provision shall remain in full force and effect.